D1070431

Pete VAN EVERA
1995

THE DIVIDED CIRCLE

THE DIVIDED CIRCLE

A History of Instruments for Astronomy, Navigation and Surveying

J. A. Bennett

Phaidon · Christie's
Oxford

To France

Phaidon · Christie's Limited,
Littlegate House,
St Ebbe's Street,
Oxford OX1 1SQ

First published 1987

British Library Cataloguing in Publication Data

Bennett, J. A.
The divided circle: a history of instruments for astronomy, navigation and surveying.
—— (A Phaidon · Christie's collectors monograph)
1. Measuring instruments——History

I. Title
620′.004 QC100.5

ISBN 0 7148 8038 8

Designed by Tim Higgins

Printed and bound in England by
Butler and Tanner Limited, Frome, Somerset

Frontispiece Reflecting circle on a stand by Troughton of London, repeating circle by Secretan of Paris, and a surveying circle by Jacobus de Steur.

Photograph credits

I am grateful to the following museums for permission to reproduce photographs of instruments in their collections (the photographs are indentified by the locations given in the captions): Bologna University Observatory Collection; Deutsches Museum, Munich; Harvard University Collection of Historical Scientific Instruments; Istituo e Museo di Storia della Scienza, Florence; National Maritime Museum, London; Trustees of the National Museums of Scotland (location: Edinburgh, Royal Museum of Scotland); Musée d'Histoire des Sciences, Geneva; Museum of the History of Science, Oxford; Nederlands Scheepvaart Museum, Amsterdam; St Andrew's University, Department of Physics Collection; Trustees of the Science Museum, London; Whipple Museum of the History of Science, Cambridge. Photographs of instruments without locations are reproduced courtesy of Christie's. I am grateful also to the British Library Board for permission to reproduce fig. 7, to the Syndics of Cambridge University Library for figs 8, 10, 18, 34, 35, 40, 53, 54, 71, and to the Museum of the History of Science, Oxford for fig. 129. Other illustrations are from the Whipple Library, Cambridge.

Contents

Preface

The structure of the book is based on three historical periods: from ancient to early modern times (Chapters 1–5), the eighteenth century (Chapters 6–9) and the nineteenth century (Chapters 10–13). A short, concluding chapter (14) deals with the twentieth century. Within each of the major divisions there are chapters on astronomy (1, 7, 11), navigation (2, 8, 12), surveying (3, 9, 13) and the instrument-making trade (5, 6, 10). The only exception is Chaper 4, which deals with the important general developments of the seventeenth century.

I have been helped by many museum curators, particularly in acquiring photographs, and I am grateful to Margarida Archinard (Geneva), Alessandro Braccesi and Giorgio Dragoni (Bologna), Alto Brachner (Munich), Paolo Brenni (Florence), Owen Gingerich and Ebenezer Gay (Harvard), Willem Mörzer-Bruyns (Amsterdam), Allen Simpson (Edinburgh), Carole Stott (Greenwich), Gerard Turner (Oxford) and Michael Wray (St Andrews). The omission of a representative of the Science Museum, London, merely indicates the efficiency of their public ordering service, and I made use of a similar service at the National Maritime Museum, Greenwich. I am also grateful to Jeremy Collins for locating photographs preserved at Christie's, and to Don Manning for taking the photographs of instruments at the Whipple Museum.

1

Foundations in Astronomy

STRONOMY, NAVIGATION AND SURVEYing were for long the three principal domains of the practical mathematical sciences, and their instruments represented contemporary standards of precision and accuracy. The central instruments of each field were designed for measuring angles between distant objects, and so, while different types were adapted to individual functions, there is a broad affinity over the entire range. At their core each of the principal instruments – whether quadrant, sextant or theodolite – depended on a sight moving over the whole or part of a divided circle, a circle or arc carrying a scale, generally marked in degrees.

The makers of these instruments, especially of the astronomical instruments, were the élite of the trade. They made possible the ambitious observational programmes of the astronomers. They built classic observatory instruments that achieved heroic status in the history of science. They set goals in precision for the rest of the trade and, through instruments of navigation and surveying, dispensed new standards of accuracy in practical problems.

Their instruments are enjoyed at first in aesthetic terms. Made in brass, mahogany, ivory and silver, their elegance is based on symmetry, simplicity of line, attention to detail and restrained decoration. But they are more than beautiful objects: they have conceptual and practical functions. They rest on complex mathematical techniques, founded on rarefied theories of geometry and astronomy. Yet the great majority of surviving instruments were used for such everyday necessities as position-finding at sea, or canal or railway survey. The whole enterprise depended on a conviction that technological problems could be managed by mathematical principles, a conviction which the instrument could focus for an instant on the solution of an immediate practical problem.

Astronomy and the division of the circle

Throughout the history of the precision instruments of the mathematical sciences, astronomy represents the leading edge of the discipline's development. It is always in astronomy that the highest standards of accuracy and instrumentation are found, as well as the most prestigious commissions, whose successful accomplishment might establish a maker's reputation. It is also from astronomy that many of the working concepts are derived.

Astronomical phenomena regulate the most basic ordering of life's routine, from day and night and seasonal change at the more primitive level, to the changing visibility of the constant star patterns, the phases and motion of the Moon and the visibility of the planets at the more sophisticated. Even the most primitive organized communities need to relate astronomical events to the agricultural cycle, or to the life cycles of animals. Civilizations and centralized administrations soon require a caste of specialists, concerned with calendaric regulation and the principal rites of state and religion. The set of techniques used to relate astronomical to earthly events can readily develop into a complex and sophisticated science, designed to meet the demand for a full account of how the stars and planets influence the patterns of our lives.

The approximate coincidence between the number of days in the year and the number of degrees conventionally ascribed to the circle might seem to suggest an astronomical origin for our most common division of the circle. If this were so, it would derive ultimately from the number of times the Earth rotates during one entire revolution around the Sun, and in the old geocentric astronomy a degree would then be represented by the angular distance travelled by the Sun during one day in its annual passage through the zodiac. It seems, however, that the approximate coincidence between days in the year and degrees in the circle does not precisely explain the origin of the degree, but may have assisted the general adoption of this conventional division.

The 360–degree division originated in the astronomy of the Babylonians, but their civil calendar was based on the lunar and not on the solar cycle. Its origin is plausibly linked to the 'sexagesimal' system of Babylonian computation (established considerably before the degree), with six repetitions of 60 units comprising a circle.

Hellenistic astronomers adopted the notion of 360° to the circle from the Babylonians, whose astronomy became an important resource for the development of their science, and it was the work of the most famous Hellenistic astronomer, Claudius Ptolemaeus (Ptolemy) (AD *c.*100 – *c.*170), that established the methods and reference of astronomy for many centuries to come. They also adopted, for calendaric purposes, the much more practical system of the Egyptians, where 12 months each contained 30 days, making an annual total of 360, plus an additional 5 days added at the end of each year. This was originally a purely practical, civil calendar, related to the agricultural cycle and not to astronomy (there was also a religious calendar based on the Moon), but the importance of the annual inundation of the land by the Nile, heralded by the first appearance of Sirius, linked it to the solar year.

If 360° became settled as the most familiar division of the circle, others used in astronomy derived directly from the solar year and the daily cycle of the heavens. We can think of the two most common secondary divisions as related respectively to the annual and daily apparent motions of the Sun. Circular scales on astronomical instruments are sometimes found divided by date, perhaps with the embellishment of the twelve zodiacal signs, each of which is traversed by the Sun in approximately a month (though the month, of which the week is a subdivision, in fact derives from the lunar cycle). Such scales bear some relation to the apparent annual solar motion. They may, for example, be positioned, when in use, in the plane of the ecliptic (the apparent annual path of the Sun through the stars, which of course rotates daily with the celestial sphere), for measuring positions in what are called 'ecliptic co-ordinates' – celestial latitudes and longitudes.

Circles set parallel to the equator, on the other hand, are commonly divided into hours and minutes, and so relate to the apparent daily motion of the Sun. The Sun appears to move in a path parallel to such a scale, called an 'equinoctial' (equal hour) circle, since the hour divisions are equal: the shadow cast by a perpendicular line through the centre will move through equal divisions in equal times. The conventional division into 24 hours derives from Egyptian timekeeping, though they originally reckoned 12 hours between sunrise and sunset and 12 corresponding hours of night. In other words, day and night hours were of different lengths, and they varied according to the time of year. The Babylonians used equal hours and this idea was adopted for astronomical purposes, as was the Babylonian subdivision into 60 minutes. Angular position in the 'equinoctial' or 'equatorial' circle came to be expressed, as it is still, in hours, minutes and seconds, and is known as 'right ascension'. The complementary co-ordinate, angular distance above the celestial equator, is expressed in degrees and known as 'declination'.

Other specialized divisions of the circle were used in navigation and in surveying, but in astronomy the three divisions – by degree, by date and by time – were distributed, as appropriate, among the three systems of co-ordinates in which positions were measured. These were azimuth and altitude (where the observer's horizon provided the principal reference circle), right ascension and declination (referred to the celestial equator) and ecliptic longitudes and latitudes (referred to the circle of the ecliptic). As a final complication, declination scales, though properly divided by degrees, can also carry dates, because of the importance of solar declination, which varies throughout the year (a variation determined by the inclination or 'obliquity' of the ecliptic to the celestial equator) and so can be indicated by date.

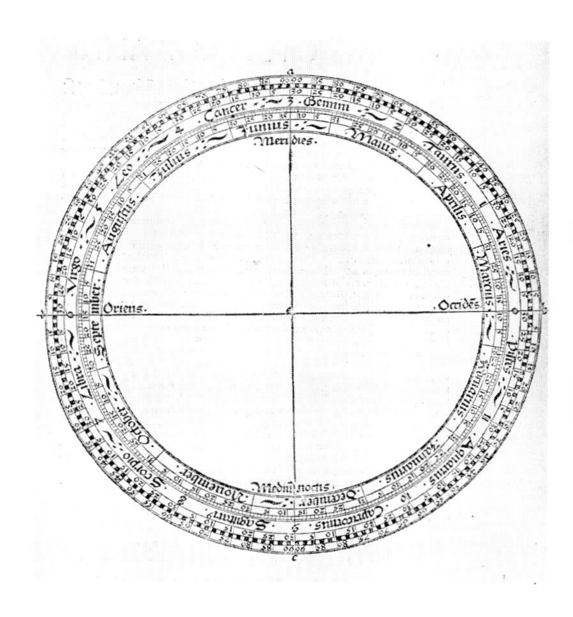

1 The circle divided by date, by signs of the zodiac and by degrees, from the account of the astrolabe by Johann Stoeffler. (Stoeffler, 1524)

The mathematical sciences

Ancient astronomy, and Hellenistic astronomy in particular, also provides a useful introduction to the nature and domain of the mathematical sciences, on which the reference of this book is based. It will surely strike the modern reader as odd that a book dealing with the instrumentation of astronomy does not cover the telescope in its own right, nor with its optical development, but only with its role as a sighting device on instruments for measurement in astronomy, navigation and surveying. The distinction would have been understood immediately by practitioners working in by far the greater part of the period covered here. When the telescope was introduced into the study of the heavens in the seventeenth century, it was not, properly speaking, an astronomical instrument at all, neither was it understood as such, for astronomy was considered a mathematical science and the telescope was not a mathematical instrument. It could not, of itself, be used for measurement, but only for making qualitative observations. Such observations – of the Moon, Sun, planets and stars – were of interest, but not within the traditional mathematical science of astronomy, which dealt principally with mathematical planetary theory. They were relevant rather to a distinct domain, which embraced the heavens within its reference and was known as 'natural philosophy'.

The source of this distinction is found in Greek philosophy and in the difference between 'material' and 'formal' accounts of the natural world. Theorists who subscribed to material accounts held that the organizing principles of nature were properly described in material terms – fire, water, atoms, etcetera. Formal theories, on the other hand, invoked principles expressed in abstract, usually mathematical, terms, such as the numerical relationships of the Pythagoreans or the geometry of Plato.

According to Plato (427BC – 348/7BC), the world was fashioned in imitation of a perfect geometrical model – a model which exists in its complete expression in a realm transcending our own, but is only imperfectly realized in our physical world. Access to the transcendental realm, which Plato regarded as the true and permanent reality, is purely intellectual, through our individual memories of our previous non-physical existences in the world of ideas. Geometry is, of course, an example of the kind of intuitive understanding that Plato characterized as remembering.

It follows from this account that the true astronomer is ultimately concerned with the perfect geometrical model, rather than with its imperfect physical expression, however useful this may be as a source for clues to the original blueprint. Astronomy becomes a mathematical science, whose goal is a geometrical model, which on the one hand accounts for our observations, and on the other exhibits characteristics appropriate to the transcendental realm: beauty, symmetry, relation and harmony, expressed through the perfection of uniform circular motion. It does not concern itself with the imperfect physical realization of this model – with the natures of planets and stars, or with the physical mechanisms that produce their motions.

When Aristotle (384BC – 322BC) came to account for the natural world, he found all this profoundly unhelpful. For him the physical world of sense experience, rather than some transcendental realm, must be the object of attention: this is the only world we can ever know, and we come to know it, not by pure thought, but by the senses. Aristotle's account was a qualitative one, in which a 'material' component was conspicuous – the four sublunary elements of earth, water, air and fire, and the fifth element or 'quintessence' of the heavens, together with the principles which governed their motions and their relations to each other. Mathematics was not central to Aristotle's understanding of the world, since, however useful it may be in describing aspects of the behaviour of material things, it did not penetrate to their true natures. His qualitative picture thus accorded with his basic assumptions about the organization of the natural world. His natural philosophy dealt with the heavens, but in qualitative terms, and in answer to just the kinds of question abandoned by the mathematical astronomers: the natures of the heavenly bodies, their physical characteristics, the causes of their motions.

If we seem to have strayed, however briefly and superficially, into metaphysics, this can readily be justified. Part of the definition of our subject – its boundaries and characteristic assumptions – has its source in the philosophical differences between Plato and Aristotle. Astronomy, as a mathematical science, came to be seen as distinct from natural philosophy, the one concerned with geometrical theories of heavenly motion, the other with physical characteristics and causal relations throughout the cosmos, heaven and Earth. The observations relevant to such a mathematical astronomy were positional determinations, and its associated instrumentation comprised measuring devices. Geometrical models were constructed so as to derive – to 'save', to use the technical expression – these observations, and such models could, of course, be used predictively, that is, to derive astronomical events yet to be.

It was in this predictive aspect – either in relation to future planetary positions (it should be remembered that the Moon and the Sun, though not, of course, the Earth, were considered planets in ancient astronomy) or past, unobserved positions – that mathematical astronomy first found a 'practical' application, and this practical element became more important as the goals of Hellenic astronomy were revised in the Hellenistic context. The practice of astrology required a predictive or retrodictive apparatus of just this type to foretell favourable or unfavourable aspects, or to construct horoscopes. Calendaric regulation in any advanced form, as well as accurate timekeeping, also required mathematical astronomy. Astronomy thus grew to fulfil a role that became closely associated with mathematical science in general: already divorced from physical or causal theorizing, it found an outlet rather through practical applications.

These early practical applications were augmented in the Renaissance, when new demands on navigational theory and practice were one result of broad social changes, which brought political and commercial importance to the practical potentials of the mathematical sciences. Navigation became the practical application most influential to the development of astronomy, and together with surveying it was part of a remarkable blossoming of practical mathematical science in sixteenth-century Europe. By this time astronomy was established as a model for the individual subjects within the domain. Based on mathematics, divorced from natural philosophy, applied to practical ends, astronomy stood as the oldest and most esteemed of the mathematical sciences, which were set for a spectacular expansion through the vigorous promotion of their enthusiastic practitioners. The characteristics of the mathematical sciences – with astronomy, navigation and surveying as the most prominent among them – were their technical basis in mathematical theory, their practical applications to useful ends, and their use of instruments. The twin goals of astronomy – planetary theory and applications to timekeeping and navigation – both required instruments designed to make angular measurements, and instruments of the same general kind served in navigation and surveying.

While instruments were the trademark of the mathematical sciences, the telescope fell outside their reference. The seventeenth century, lacking our general notion of a 'scientific instrument', had in fact to create two new categories to add to 'mathematical instruments', in order to cope with the broadening reference of newly devised instrumentation. First the term 'optical instruments' was used to classify the telescope and the microscope, and then 'instruments of natural philosophy' was invoked to deal with such radical innovations as air-pumps, barometers, thermometers and electrostatic generators. The acceptance of instruments within natural philosophy depended, of course, on a wholesale rejection of the qualitative natural philosophy of Aristotle. But the categories of mathematical, optical and natural philosophical instruments survived well into the nineteenth century, and our concern with the principal mathematical instruments is wholly consonant with how they were understood and categorized by those who made and used them.

The instruments of Ptolemy

The instruments described in Ptolemy's *Almagest* provide the clearest record of the precision instruments of Hellenistic astronomy, and they became the starting-point for the instrumentation of astronomy in western Europe, when Latin translations of Ptolemy's work became available in the twelfth century. The observatory instruments he describes were each designed for a single purpose, for making a certain kind of observation. His meridian armillary or ring, for example, was designed for measuring the altitude of the Sun when due south, i.e. when it crossed the observer's meridian at noon, a measurement known as the meridian altitude. The instrument consisted of a fixed, graduated, bronze ring, mounted in the meridian, with a close-fitting, concentric ring moving within it. The inner ring carried two plates at diametrically opposed positions, with two accompanying pointers, moving over the outer scale. A plumb-bob and a meridian line were used to align the instrument. Ptolemy describes its use as follows:

> Having set the instrument up in that way, we observed the Sun's movement towards the north and south by turning the inner ring at noon until the lower plate was completely enshadowed by the upper one. When this was the case, the tips of the pointers indicated to us the distance of the Sun from the zenith in degrees, measured along the meridian. (Ptolemy, 1984, p.62)

Thus Ptolemy used the lower plate as a target for the shadow of the upper. Writing in the fifth century, Proclus (410–85) gave a detailed account of the construction of a similar meridian ring in his *Hypotyposis astronomicarum positionum*, but the plates have holes and are thus closer to conventional sights. A different armillary instrument mentioned by Ptolemy was the more primitive equatorial ring, consisting simply of a bronze ring set in the equatorial plane. It was ungraduated, had no sights, and was used simply to determine the equinoxes, which were indicated when the shadow of the upper section of the ring fell on the inner surface of the lower.

An alternative instrument to the meridian armillary, used for making the same kind of observation, was the 'plinth', a forerunner of the later mural quadrant, but here again a shadow, rather than sights, indicated the position to be measured. A vertical plane of wood or stone was marked with the 90° of a quadrant, with each degree subdivided further, and set in the meridian. A peg at the apex or centre of the quadrant cast a shadow on the scale, which thus indicated the meridian altitude of the Sun and a plumb-bob, hung from this peg, was used to align the edge of the quadrant with the vertical.

By far the most complicated of the instruments described in the *Almagest* was the armillary sphere (fig. 2), a skeletal sphere comprising a number of interlocking rings, seven in all in the Ptolemaic instrument. The two outermost rings resemble the meridian armillary, and like it are set in the meridian. The movement of the inner ring of the pair, however, is not for making observations, but for establishing a polar axis (one which points to the celestial pole, around which the heavens appear to rotate daily) on which the remainder of the instrument may be moved. It is thus adjusted according to the geographical latitude, which is equal to the altitude of the celestial pole. The result of mounting an instrument in this way is that the motion of a star can be followed by moving the sights in only one co-ordinate – parallel to the equator – which is easily done by hand, and the principle was of considerable importance to the development of astronomical instruments.

The ring (the third from the outside) mounted on the polar axis carries a second axis at an angular distance from the first equal to the obliquity of the ecliptic. This established the innermost four rings as an ecliptic co-ordinate system, the first being used merely to carry, fixed to it at right angles, the graduated ecliptic ring itself. The innermost two rings are arranged in a manner similar to the two rings of the meridian armillary, the inner with two diametrically opposed sights, the outer graduated for ecliptic latitude. This pair of rings pivots on the second axis and so is perpendicular to the graduated ecliptic ring, to which it almost extends and around which it can rotate to measure latitudes in any ecliptic longitude.

Thus the whole instrument was designed to allow angular measurements to be taken in ecliptic co-ordinates, which were required by the geometrical methods of Ptolemy's planetary theory, and to avoid any need to convert from altitudes and azimuths. The result of this desideratum was a very complex and sophisticated instrument, whose accuracy and practicality must have been limited, but which contained the seeds of many developments to come.

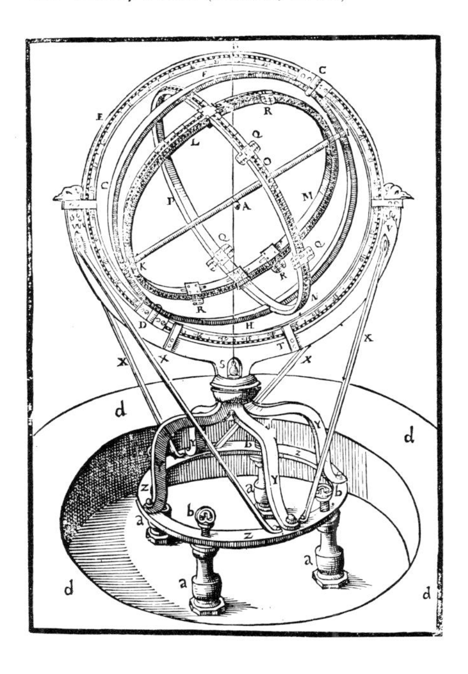

2 An armillary sphere of the Ptolemaic type, diameter 1.17m. Built long after Ptolemy's era, it belonged to the sixteenth-century astronomer Tycho Brahe. The polar axis, CD, and the ecliptic axis, IK, are clearly seen. (Brahe, 1598)

The parallactic instrument, more commonly referred to as 'Ptolemy's rulers', had the potential for versatility, and was thus the instrument most used by later astronomers, but Ptolemy described it as designed for taking the zenith distance of the Moon. Zenith distance is the angle between an object and the observer's zenith (the point directly overhead), this angle being the complement of the altitude, and the instrument's technical name comes from the fact that Ptolemy used such measurements to determine lunar parallax (differences in angular measurements deriving from the distance of the Moon from the observer). The more common name derives from the instrument's physical form.

A vertical post, adjusted with a plumb-bob, is graduated with a linear scale and carries two pivoted arms. One pivots at the foot of the post; the other has sights at either end and is attached to a pivot fixed to the top of the post. A reference point close to the free end of the upper arm is set at a distance from the top pivot equal to the graduated distance between the pivots. When the sights are aligned with the Moon, the lower arm is made to intersect the upper at the reference point, so that the distance along the lower arm is the chord subtended by the angle between the post and the upper arm, which is equal to the zenith distance. The lower arm is then pivoted up so that the chord length can be taken off the graduated scale, and the angle readily found by reference to a table of chords.

The obvious expedient of graduating the lower arm itself was adopted in later versions of Ptolemy's rulers (fig. 3), but the practical advantage of a linear scale was preserved: it avoided all the difficulties and inaccuracies associated with dividing the circle. A further important feature of this instrument was that it included an alidade – a straight rule with sights mounted at either end. In no other instrument described in the *Almagest* are the sights or target plates connected in this way, but the alidade was to become a typical feature of precision instruments until, and even considerably after, the introduction of telescopic sights.

3 Ptolemy's rulers. Like the instrument in fig. 2, this is a late example; it was used, and reputedly made, by Copernicus, and afterwards incorporated into the observatory of Tycho Brahe. (Brahe, 1598)

The alidade is also conspicuous on the most famous Ptolemaic astronomical instrument, but one which was not described in the *Almagest*, namely the astrolabe (figs 4, 5). The astrolabe was known to Ptolemy, and an even earlier history is the subject of speculation. It falls within the class of mathematical instruments, indeed it must be regarded as one of its most significant members, both on account of its conceptual complexity and because it was widely used outside what might be considered as the professional mathematical community. However, as a portable device, having a number of everyday applications, its historical role is closer to such instruments as sundials, horary quadrants and nocturnals, which were generally useful and were produced in considerable numbers, but whose development was related only tangentially to that of the precision instruments of the professionals and specialist scholars. Such instruments will be covered here, but also tangentially; they properly fall outside the central account, and in any case are already supplied with a considerable secondary literature.

Physically, the astrolabe consists of a series of circular plates and straight rules held together by a removable, central pin. The most substantial member of this ensemble is a plate, thicker than the others, and provided on one side (generally regarded as the front) with a raised edge, running round the entire circumference. This plate is called the 'mater'. The back is equipped with a number of scales, including an outer circle divided into degrees, and a centrally pivoted alidade with sights. A suspension ring, fixed by a pivot to the rim of the plate, coincides with the 90–degree mark, from which the scale runs to zero, to 90°, to zero and back to 90°. Thus, when the instrument is suspended, the alidade can be used to measure altitudes, either of celestial bodies or of distant objects on earth.

On the front of the mater, the raised circumference (known as the limb) carries a degree scale and often a scale of hours, and the hollowed centre houses a series of alternative plates, which are generally fixed in position by lugs locating a hole in the limb. Each side of these plates is engraved with a network of lines, which represent a celestial co-ordinate system of altitudes and azimuths projected onto the plane of the equator. Thus the celestial pole is represented by the centre of the plate, where there is a hole for the pin, and the observer's zenith is a point on the vertical meridian line, at a distance from the pole equivalent on the projection to the complement of the latitude for which the plate has been drawn. Around this zenith point cluster lines of equal azimuths and from it radiate lines of equal altitudes. When the observer selects the plate appropriate to his latitude and fits it into the mater, he establishes a reference system of celestial co-ordinates based on his own horizon.

Above the latitude plate is a skeletal plate known as the 'rete', which is essentially a star map, derived from the same planispheric

4 (*left*) Astrolabe by Georg Hartmann of Nuremberg, 1532, diameter 137mm. The front shows the rete, a pivoted rule and degree and hour scales on the limb; the rete has pointers for 27 stars and an ecliptic ring divided to 1 degree. Beneath the rete is a latitude plate; there are three such plates, each engraved on both sides, for latitudes between 39° and 54° north.

5 (*right*) The reverse side of the astrolabe illustrated in fig. 4 There are engraved degree and calendaric scales, lines for finding the time in 'unequal hours' (where day and night are divided into exactly 12 hours throughout the year) from information on solar altitude, and a shadow square. There is also a pivoted alidade with sights for making observations.

projection of the celestial globe. Prominent stars are represented by the tips of pointers, and the ecliptic by a band engraved with the signs and degrees of the zodiac. The celestial equator is sometimes represented in whole or in part by a band concentric with the central pole. Unlike the latitude plates, the rete is free to rotate about the pin, and this rotation represents the daily apparent motion of the celestial sphere, with the changing positions of the stars in altitude and azimuth given by the network of lines visible beneath.

Thus the astrolabe is a movable model of the heavens, which can be set to their present configuration or to any of special interest (such as that at a time of birth). To establish the present configuration, for example, it is sufficient to measure the altitude of one of the stars represented on the rete, by using the alidade and degree scale on the back, and then to set the star's pointer on the appropriate altitude line. The time, as well as positions of other bodies, can be read off immediately. By using the zodiac calendar (a scale relating date to the Sun's position in the zodiac), generally engraved on the back of the astrolabe, in conjunction with the ecliptic band on the rete, times of sunrise and sunset for the given latitude can be calculated for any date. Times of rising and setting of stars can also be found, and a number of other calculations performed. A rule, which pivots in front of the rete, is used to relate the central projections to the scales on the limb. A number of different additional scales may be found on the back, such as the 'shadow square', which is used in surveying.

The complexity and sophistication of the astrolabe makes any explanation, even in outline, a long-winded business, but the space it must command is not a reflection of its importance in the technical development of precision instruments. It was significant in two related respects: its use fostered mathematical and astronomical literacy, and its commercial potential was important in the establishment of centres of mathematical instrument-making. But for the instrumentation of precision astronomy, the relatively straightforward tools described in the *Almagest* became the vital resource, whose features included meridian and equatorial instruments, the use of the alidade, of a polar axis and of measurement in ecliptic co-ordinates.

The Ptolemaic tradition

Ptolemaic astronomy was preserved and fostered by Islamic scholars, after the expansion of Islam's political influence from the seventh century onwards had brought them into contact with Greek learning. A number of observatories were built and supplied with large, fixed instruments. A notable example was the observatory built at Bagdad in the ninth century, under the patronage of the Caliph al-Ma'mum, and equipped in the Ptolemaic tradition. An example of a private observatory of this period was that of the famous astronomer al-Battani (d.929), whose instruments included an armillary sphere, a mural quadrant (a quadrant fixed to a meridian wall) and Ptolemy's rulers.

The greatest Islamic observatory was certainly that founded at Maragha in 1259, under the direction of Nasir al-Din (1201–74), and impressively furnished with large instruments. Among these was a mural quadrant with a radius of 4.3 m; it was made of hardwood with a copper limb – the arc on which the graduations, in this case to minutes, were engraved. There was a meridian ring of 2.5 m diameter and an armillary sphere with a diameter of 3.2 m, together with large examples of Ptolemy's rulers. Each of the instruments mentioned here was equipped with an alidade. Perhaps the most original instrument was an azimuth ring furnished with two quadrants that could be rotated above it to any azimuth, i.e. 'altazimuth' (altitude and azimuth) quadrants, capable of simultaneous measurements of both co-ordinates. The provision of two quadrants, with alidades, on one ring allowed azimuth differences to be taken directly.

The Islamic tradition thus made significant improvements in Ptolemaic instrumentation, especially in the general adoption of the alidade; significantly 'alidade' is the Arabic name by which the device is always known. The importance of large instruments was also appreciated. The final monument to this phase of Islamic astronomy paid exaggerated respect to size. The observatory of Ulugh Beg (1394–1449), built at Samarkand in the fifteenth century, had enormous stone instruments, including a meridian sextant (an instrument with an arc of approximately 60°) 40 m in radius. On an instrument of this size degree divisions were 0.7 m long and each millimetre represented five seconds of arc. By this time, however, astronomers in western Europe were also engaged in Ptolemaic astronomy, and had taken over the initiative for its development from the Islamic world.

Muslim Spain, with its observatory at Toledo, had been an important channel for astronomical communication between east and west. By the second half of the twelfth century Toledo had become the most active centre for translation of scientific and medical works from Arabic into Latin, to satisfy the growing demand of western scholars. The most famous translator at Toledo was the Italian Gerard of Cremona (*c.*1114–87), and it was at Toledo that he discovered Ptolemy's *Almagest* in Arabic and made a complete translation into Latin. The *Almagest* was only one instance of Gerard's enormous output in translation, but it was one of the most significant.

The 'torquetum' or 'turketum' (the 'Turkish' or 'Muslim' instrument, fig. 6) is characteristic of medieval European astronomy, and demonstrates the Ptolemaic intellectual heritage as well as the debt to Islamic invention. Latin accounts of it can be traced to the thirteenth century and a treatise by Franco of Polonia. There are many subsequent descriptions and variants, and a whole host of 'universal' instruments are its successors, down to the 'portable observatories' of the early nineteenth century.

The torquetum is equipped to take measurements in all three sets of astronomical co-ordinates, and to convert readily from one to another, a desideratum of Ptolemaic planetary theory. Two of the best-known accounts are by the later astronomers Regiomontanus (1436–76) and Apianus (1497–1552), and the instrument illustrated here is that of Apianus. While details differ between various descriptions, the principles are the same.

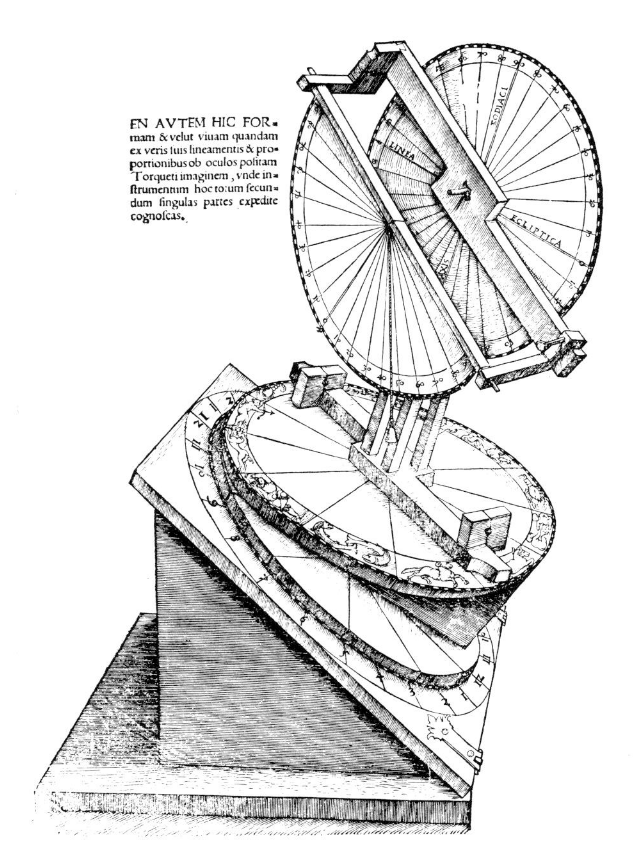

6 (*left*) The folio-sized woodcut prepared to accompany Peter Apianus' account of the torquetum, where he describes its assembly stage by stage. (Werner, 1533)

7 Richard of Wallingford dividing a circular instrument. He is using a pair of dividers to mark out a circular plate, while there are other tools on the bench and a quadrant hangs to one side. Richard's face is spotted, to indicate that he suffered from 'leprosy'. British Library Ms Cotton Claudius E.iv, f.201.

A pair of folding plates are connected by hinges. One rests level, while the other is propped at an angle to the first equal to the complement of the latitude (the co-latitude). This has the effect of setting the upper plate parallel to the equator, i.e. locating the equinoctial or equatorial plane. Apianus' instrument is made for a single latitude, but it was possible to have the inclination of the upper plate adjustable, so that the instrument could be used in different latitudes. The circle on the upper plate is divided by hours, since it is parallel to the equator and the remainder of the instrument moves in right ascension as it rotates above it.

The circle above the upper plate is propped at an angle to it equal to the obliquity of the ecliptic, so that it can be rotated into the plane of the ecliptic. This circle is thus divided by degrees and signs of the zodiac, and the alidade rotating on its centre measures celestial longitudes. On this alidade is mounted, at right angles to the ecliptic circle, a circle and alidade for taking celestial latitudes. An additional feature of the Apianus torquetum is the semicircle attached to the upper alidade. This is equipped with a plumb-line, which of course defines the zenith, so that it indicates a body's altitude as the alidade indicates the celestial latitude.

The torquetum can be adapted to configurations alternative to the standard one. With the ecliptic circle folded flat on the equatorial plane, measurements may be taken in right ascension and declination. When the two base plates are also folded, the instrument measures azimuths and altitudes.

Other novel instruments sprang from medieval concern with Ptolemaic astronomy. The 'rectangulus' of the English astronomer Richard of Wallingford (*c.*1292–1336) performed the same role as the torquetum, but by using a succession of pivoted rods instead of circles. As with Ptolemy's rulers, straight scales were more easily divided, but the rectangulus enjoyed no general popularity.

The cross-staff of Levi ben Gerson (1288–1344), which has the same advantage as the rectangulus, proved to be far more popular. In its most basic form, the cross-staff (fig. 26) consists of a straight wooden rod of square cross-section (the 'staff', 'radius' or 'index') which fits snugly through a central hole in a wooden cross-piece (the 'cross' or 'transversarius'). One end of the staff is held close to the eye, and the ends of the cross subtend different angles at the eye as it is moved along the staff. The staff is graduated accordingly and measurements are taken by positioning the cross so that its ends coincide with the objects whose angular distance is required. A large instrument, for astronomical use, may be mounted on a stand.

The versatility of the cross-staff is obvious: it can be used to measure angles between any two positions and is not confined to measuring a particular co-ordinate. It can be made in very different sizes, appropriate to astronomical, surveying or navigational use. Crosses of different lengths can be provided, suitable for different ranges of angles, and corresponding scales engraved on different sides of the staff. A number of precursors to Levi's cross-staff have been claimed in instruments of similar or related type (he himself had several different forms), and we shall see that there were certainly a great many successors.

The development of instrumentation within what we have called 'the Ptolemaic tradition' was thus not restricted by the tools described in the *Almagest*, even though the theoretical aims of the practitioners were bounded by Ptolemy's astronomy. By the later fifteenth century, however, a new impetus and a sense of change can be detected among European astronomers. Although theory remained Ptolemaic for a time, they were not immune to the intellectual confidence of the Renaissance, and a new vitality in astronomical practice was eventually joined by the proposal of a radically different cosmology.

Renaissance astronomy

The astronomical partnership between Georg Peurbach (1423–61), who taught astronomy at the University of Vienna, and his sometime pupil Johannes Müller, known as Regiomontanus, may perhaps be taken as an early instance of a new confidence in the pursuit and promotion of astronomy. In the 1450s they collaborated in an attempt to raise the general standard of astronomical teaching, and their reform was cast in the humanist mould of a return to classical sources, just as that of Copernicus would be in the following century. After Peurbach's death, Regiomontanus travelled in Italy and elsewhere, before settling in Nuremberg in 1471. Here he established a private observatory, convinced of the importance of a new observational basis for the restoration of astronomy. He wrote on astronomy and instrumentation, and having installed a printing press at his house, printed and published many mathematical and astronomical

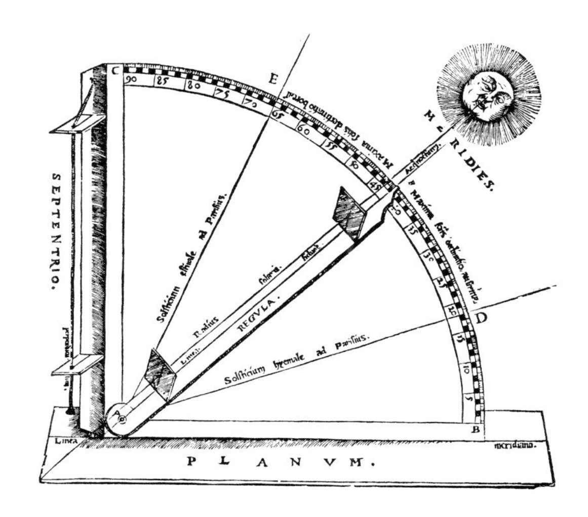

8 The astronomical quadrant illustrated by Oronce Fine, set in the meridian and leveled by a plumb-bob. The limb is divided to degrees, and each degree subdivided further. The alidade ('regula') pivots at the eye end. (Fine, 1542)

works. He was also engaged in the design and development of instruments, and his activities helped to make Nuremberg a centre for the instrument-making trade.

Peurbach had written an account of an instrument known as the geometrical quadrant (fig. 36), which consisted of a square – in Peurbach's case in the form of an open wooden frame – with an alidade pivoted at one corner. The two sides of the square opposite the pivot were divided into equal parts, numbered 0 to 1200 to 0 in Peurbach's instrument and divided to 20. Angles were thus measured, not directly in degrees, but as ratios or tangents, with the practical advantage that only straight lines and not arcs needed to be graduated. While Peurbach described astronomical applications, such an instrument was in fact better suited to surveying, where ratios can be employed directly, without converting to degrees. Further reference to this and such related instruments as the geometrical square and shadow square will thus appear under surveying.

The observatory of Regiomontanus had large examples of the cross-staff and Ptolemy's rulers. His published work included accounts of the torquetum and the armillary sphere, as well as smaller portable instruments such as sundials. He was involved with the manufacture of instruments and established a workshop in Nuremberg. Above all, through his deliberate promotion and dissemination of mathematical science and its instrumentation, he breathed new vigour and confidence into its development, and demonstrated that attitudes characteristic of other branches of the 'science' (knowledge) and 'art' (its application) of the Renaissance were established in astronomy.

An early beneficiary of Regiomontanus' intellectual legacy was Johannes Werner (1468–1522), born at Nuremberg, who settled there in 1498 after study and travel elsewhere. Like Regiomontanus, he made instruments himself, as well as writing tracts on their use. We have already encountered the treatise on the torquetum by Peter Apian (Apianus), which was first published in 1533 as a supplement to his edition of works by Werner.

Apianus fits a now-established pattern of being centrally concerned with instrumentation and with promoting practical astronomy. He was also interested in navigation and surveying. He designed instruments and published accounts of them, including a quadrant with applications in astronomy, horology and surveying. His major influence lay in popularizing astronomy. The elaborate and beautiful *Astronomicum Caesareum* was lavishly equipped with coloured volvelles – each movable diagram an astronomical instrument in itself – and the frequently reprinted *Cosmographia* (1524) was for long the standard popular work on astronomy, navigation and surveying.

Credit for the popularity of the *Cosmographia* does not belong entirely to Apianus. In 1529 an edition was published at Antwerp, prepared by Gemma Frisius (1508–55), and it was this version, and subsequent editions with additional material by Gemma, that became standard. Gemma, a native of Friesland (as his name indicates), studied at Louvain, subsequently practising medicine there

11 (*below*) The title-page from William Cuningham's *Cosmographical glasse* of 1559, based on the idea of a domain of related mathematical sciences. A modified version of the same title-page was used for the first English translation of Euclid in 1570 The instruments include the geometrical square, the horary quadrant, the cross-staff and the universal astrolabe. (Cuningham, 1559)

VSVS ANNVLI
ASTRONOMICI PER
Gemmam Phrysium.

MODIS OMNIBVS ORNATISSIMO
Ac vere Nobili Domino Ioanni Khreutter,
Sereniſſimę Reginę Hungarię Secretario
Gemma Phryſius S. D.

INter multa variaq; animantium genera, quæ diuerſiſſimis ac admiratione dignis effinxit natura dotibus, vix inuenias vir ornatiſſ. aliquod, quod minus ſuo fungatur officio atq; humanum genus. Quod quum a Deo Opt. Max. creatum ſit perfectiſſimum, ratione illa diuina animi parte præditum, qua & ea quæ recte fiunt eligeret, ſectaretur q;, & ea quæ præter officiū ſunt fugeret deteſtareturq;, nihil minus agit, imo quaſi quadam animi

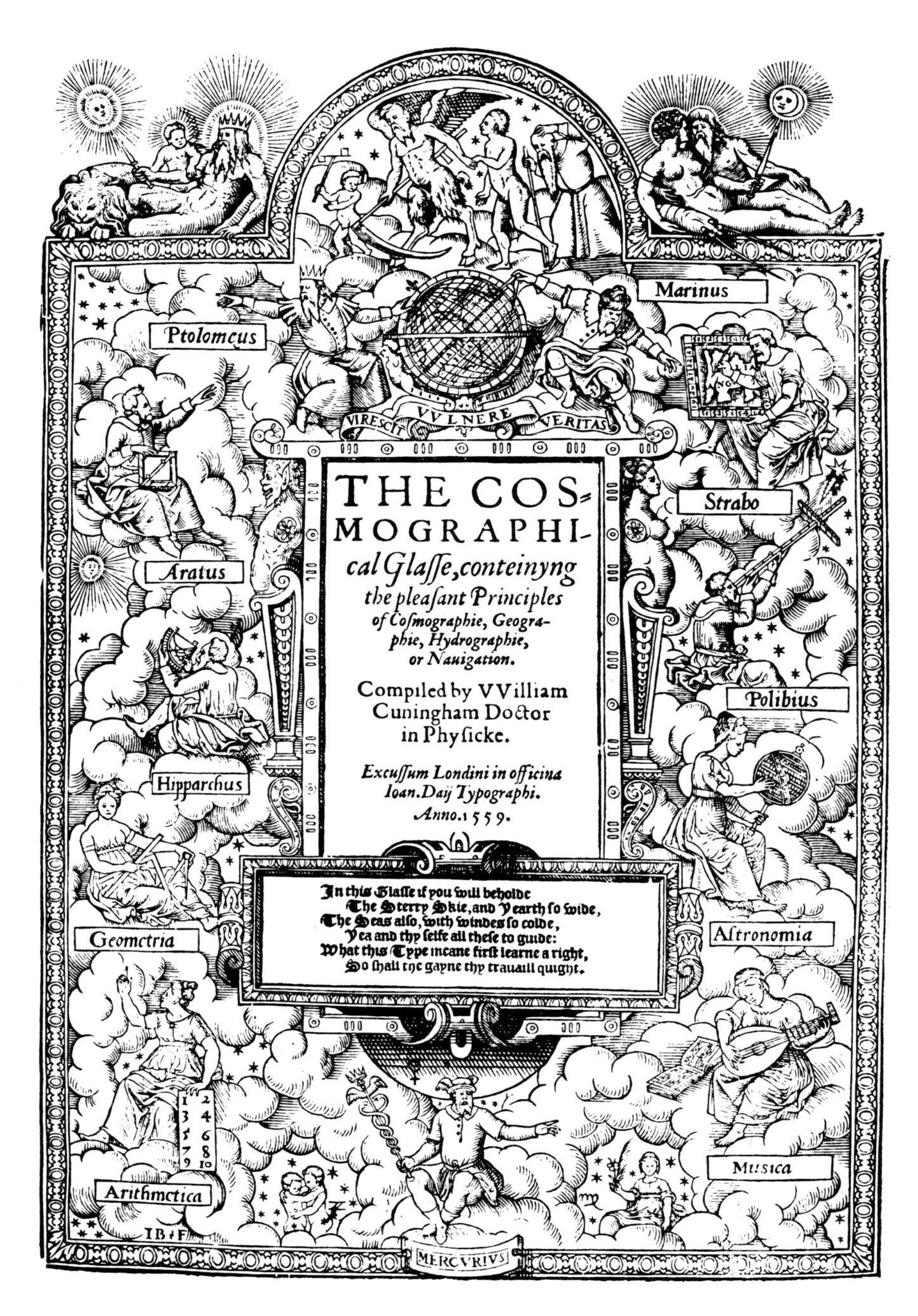

VIRESCIT VVLNERE VERITAS

THE COSMOGRAPHICAL Glaſſe, conteinyng the pleaſant Principles of Coſmographie, Geographie, Hydrographie, or Nauigation.

Compiled by VVilliam Cuningham Doctor in Phyſicke.

Excuſſum Londini in officina Ioan. Daij Typographi. Anno. 1559.

In this Glaſſe if you will beholde
The Sterry Skie, and yearth ſo wide,
The Seas alſo, with windes ſo colde,
Yea and thy ſelfe all theſe to guide:
What this Type meane firſt learne a right,
So ſhall thou gayne thy trauaill quight.

9 (*above left*) Astronomical ring dial, unsigned, 17th–18th century, diameter 93mm. An instrument whose primary function was probably for telling the time, closely modeled on Gemma's rings (fig. 10).

10 (*above*) The astronomer's rings of Gemma Frisius. A portable equatorial instrument used in any latitude (the latitude adjustment at the top is shown) for finding the time from the Sun and for a variety of approximate astronomical observations. (Apianus, 1539)

and pursuing a career as a mathematical scientist with strong interests in practical mathematics and instrumentation. He designed astronomical instruments, wrote books, such as a comprehensive account of the cross-staff, and through his influence a new centre for instrument making was formed. Probably his most celebrated protégé was Gerard Mercator (1512–94), famous for the geographical projection that bears his name, but also a maker of armillary spheres, astrolabes and other astronomical instruments. Born in Flanders, Mercator spent over twenty years in Louvain before settling in Duisburg as cosmographer to the Duke of Cleves.

Lines of influence are easily traced in the small world of sixteenth-century astronomy. Wilhelm IV (1532–92), Landgrave of Hesse, was taught mathematics by Mercator's son Rumold, and was drawn to astronomy by reading Apianus. Wilhelm established a lavishly equipped private observatory at Kassel with the aim of producing a new star catalogue. The best-known of his instruments was a substantial altazimuth quadrant, but perhaps the most significant feature of his observatory was the use of mechanical clocks. He obtained the services of the clock- and instrument-maker Joost Bürgi (1552–1632) in his efforts to upgrade his instrumentation. The improvement of clockwork from this period on was both stimulated by and vital to the development of astronomy and navigation.

Gemma Frisius and Mercator, from the north of Europe, illustrate the growth, in geographical terms, of interest in practical mathematical science. In sixteenth-century Paris also, Oronce Fine (1494–1555) was writing books on astronomy and surveying, in which instruments were prominent, as well as relatively popular works on mathematics in general and on cosmology. Popularization is a very noticeable feature of the growing volume of mathematical books of this genre, as is also what we would regard as public relations. Authors tried to spread the conviction that practical mathematics was a vital resource, which should underpin contemporary expansion in commerce, trade and exploration.

Popularization and public relations were appropriate, because a growing audience for mathematics was formed by practical men – surveyors, navigators, instrument-makers and their clients, as well as engineers, architects, practitioners of the military arts and adventurers with ambitions in exploration and colonization. Mathematical scientists sought to expand the reference of their work into other related fields, to demonstrate the relevance of their techniques and their instruments beyond their original applications, most notably into navigation and surveying. Astronomy stood as the established science, whose techniques might be applied elsewhere.

These characteristics are most obvious in the introduction of practical mathematics into England in the mid-sixteenth century – its introduction, that is to say, as a popular movement and not the isolated interest of a few individuals. Robert Recorde (*c.*1510–58) wrote popular books on arithmetic, geometry and astronomy, and was followed by Leonard (*c.*1520–59) and Thomas Digges (1546–95), father and son, who wrote on surveying and astronomy. John Dee (1527–1608) was an influential advocate of the mathematical sciences, whose activities embraced mathematics in general, astronomy and navigation.

Finding little mathematical learning in England, Dee travelled abroad between 1547 and 1551, 'to speak and confer with some learned men, and chiefly Mathematicians' (Dee, 1726, p.501). He went to Paris and to Louvain, where he spent two years with Gemma Frisius and Mercator, returning with instruments, including a cross-staff after the design by Gemma. The cross-staff was later improved by Dee's own pupil Thomas Digges.

The books of the new movement in England, though as yet envisaged only as textbooks, could be strikingly novel. Recorde's astronomy, *The castle of knowledge*, of 1556, comments favourably on the Copernican planetary scheme, published only thirteen years earlier. In 1576 Thomas Digges went further by publishing a Copernican tract, a part-translation of, and part-commentary on the first book of Copernicus' *De revolutionibus*, accompanied by a startlingly radical diagram of the universe, with the stars (which Copernicus considered fixed to a finite sphere) stretching out indefinitely into space. By the close of the century English mathematical science had produced at least two mathematicians of the first rank, Edward Wright (1561–1615) in navigation, and Thomas Harriot (*c*.1560–1621) in navigation and astronomy.

Our account of the Renaissance astronomers, informed by the perspective of instrumentation rather than theory, may seem eccentric; it is certainly not standard. It is useful here, however, since – as evidence of the growing reference of practical mathematical science – almost all of those mentioned had interests beyond astronomy, in navigation and surveying, and so appear again in later chapters. One 'great name' must, however, figure prominently even here, and although he had experience of surveying, his calling was very positively to astronomy. Tycho Brahe (1546–1601) actually carried through a programme of the kind envisaged at Nuremberg and Kassel – the provision of the first comprehensive observational base since ancient times – and in doing so he brought the tradition of the open-sighted (i.e. non-telescopic) instrument to near-perfection.

The instruments of Tycho Brahe

Tycho Brahe's famous observatory on the island of Hven in the Danish sound was established in 1576, under the patronage of Frederick II, King of Denmark. Tycho had studied and travelled widely in Europe, buying instruments, visiting mathematicians and making astronomical observations. He lived for a time in Augsburg, for example, which, with Nuremberg, had become a second centre for instrument making. He also visited Kassel, and Wilhelm's subsequent recommendation may have been influential in the king's decision to grant Tycho the use and income of Hven.

During this early period Tycho became familiar with the standard astronomical instruments: in addition to using the instruments of others, he acquired various quadrants and sextants, as well as a cross-staff, 'made', he tells us, 'according to the direction of Gemma

12 Tycho Brahe's famous mural quadrant, radius 2m, built in 1582 and divided by transversals. An assistant is shown making the observation, while another notes it down and a third calls out the time shown by the clocks. Tycho himself and one of his dogs are represented in the mural painting, and there are four instruments in the top two arches: an azimuth quadrant, an armillary sphere, Ptolemy's rulers (once the property of Copernicus) and the sextant on a ball-and-socket joint. Beneath is illustrated a study with a celestial globe and assistants at their calculations; in the basement is a chemical laboratory. (Brahe, 1598)

Frisius' (Brahe, 1946, p.108). By the time he settled in Hven his experience and vision were sufficient to plan a comprehensive and self-sufficient astronomical institution, and to begin to build a series of, eventually, some two dozen instruments. Uraniborg ('Heavenly Castle'), as he named the observatory, housed not only the instruments and a number of assistant observers, but a workshop for Tycho's instrument-makers and a paper-mill and printing press.

Of the instruments themselves, the most famous is the mural quadrant (fig. 12), a polished quadrant of brass, about 8 cm wide and 3 cm thick, with a radius of about 2 m, fastened to a meridian wall. It was divided to degrees, and subdivided further to ten seconds of arc by transversals. Tycho claimed that readings to five seconds could be taken 'without difficulty' (*ibid*, p.29).

Division by transversals was a relatively well-known technique in the later sixteenth century. It had been described by Thomas Digges, for example, in connection with his cross-staff, and there are other, scattered examples of its use. Two lines – parallel on a straight scale, concentric on an arc – are drawn at a convenient distance apart, and are divided in the same way. Diagonal lines are then drawn from each division on one line to the next division on the other. These diagonal lines, being much longer than the primary divisions, can be more easily subdivided, either individually or by means of parallel or concentric lines between the two original ones.

A large, stable, accurately divided arc was perhaps the major reason for the success of Tycho's quadrant, but there were other sophistications. A common alidade with sights introduces two errors, eccentricity and parallax: wear on the bearings may shift the centre of motion from the centre of the arc, and there is a range of possible alignments on any star with pin-holes which must in practice be of a finite size. Tycho dispensed with the alidade and substituted a foresight – a brass cylinder set in an aperture – and a backsight sliding on the limb. To position the latter sight, two parallel slits had both to be aligned with opposite edges of the cylinder, thus eliminating the parallax error. Clocks were provided for noting the times of the altitude measurements.

Tycho's great quadrant was not only a scientific instrument, but also a work of art. Paintings were added to the wall, 'for the sake of ornament, and in order that the space in the middle should not be empty and useless' (*ibid*, p.30). These included a full-length portrait of Tycho, by an artist from Augsburg, and paintings of instruments, of the general activity of the observatory and of the castle's chemical laboratories. Even Tycho's favourite dog, one 'exceptionally faithful and sagacious', was included.

Of the other instruments, some were related fairly closely to Ptolemaic precedents. There was a series of armillary instruments: three were equatorial armillaries and one was zodiacal (fig. 2), that is to say the two final rings were used, as was the case with Ptolemy's armillary sphere, for measurement in ecliptic co-ordinates. Tycho's explanation of the practical distinction between an instrument of this type and the more basic altazimuth instruments is worth noting:

> Since . . . the use of these instruments [the altazimuth type] for astronomical purposes essentially requires trigonometrical calculations which are not easily comprehensible to everybody and particularly cumbersome to certain people who shun labour, certain other appliances have been invented, with the aid of which the latitudes and longitudes of the stars, the two quantities particularly required, can be found with little inconvenience and without troublesome calculations. I find that two of these in particular were used by the ancients. One is the so-called armillary instrument that was used by Hipparchus and Ptolemy, who gave it its name. The other is called the torquetum, an instrument which in my opinion was invented by the Arabians or the Chaldaeans, and used by them. (Brahe, 1946, p.53)

The other Ptolemaic instruments at Uraniborg were two examples of Ptolemy's rulers: one (fig. 3) 'which previously belonged to the incomparable Copernicus, and was even, it was said, made by him with his own hand' (*ibid*, p.45), and a much larger version of Tycho's own devising.

Among the remaining instruments were those which derived from

13 Tycho's altazimuth quadrant, radius 1.55m. Azimuths are taken from the circle at the level of the top of the instrument, as it moves about the vertical axis. The alidade has a cylindrical foresight and two parallel near sights, so as to correct for parallax by the technique used with the mural quadrant. Tycho claimed an accuracy of a quarter of a minute for this instrument. (Brahe, 1598)

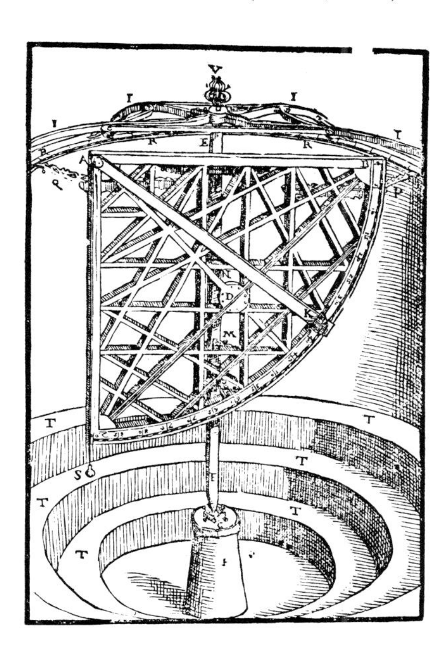

14 (*left*) An altitude sextant, radius 1.55m, described by Tycho as 'invented by myself'. He had to leave the island of Hven after the death of his patron Frederick II, and the *Mechanica* was published in 1598, the year after the final observation at Uraniborg. It was dedicated to Emperor Rudolph II, from whom Tycho hoped for new patronage. He wrote, somewhat poignantly, of this sextant that it had the advantage of being portable: 'An astronomer . . . has to be a citizen of the world, and consider every place to which circumstances or necessity might lead him as his native country'. (Brahe, 1598)

15 (*right*) A sextant of Tycho's, radius 1.55m, moveable on a ball-and-socket joint, and used by two observers to measure angular distance between any two objects. One used the fixed sights C and F, the other those on the alidade pivoted at A. Tycho had a similar instrument with two alidades. The type of cross-bracing on this instrument, the altazimuth quadrant (fig. 13) and the altitude sextant (fig. 14) would become standard for many years to come. (Brahe, 1598)

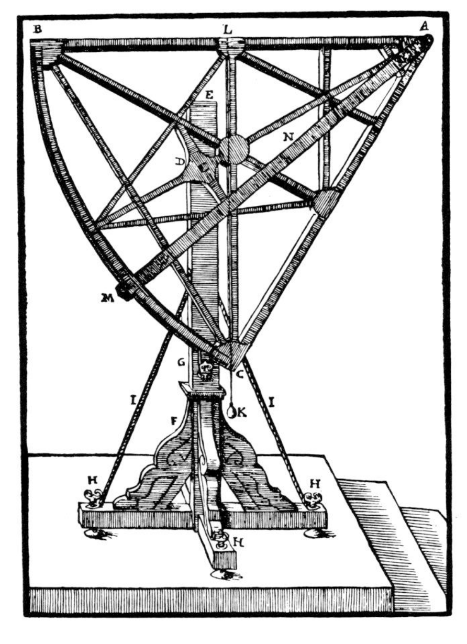

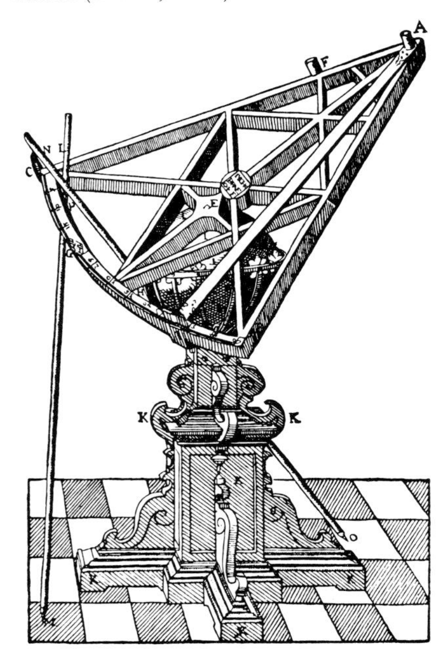

16 (*below*) Tycho's small altitude quadrant, radius 390mm, subdivided by the 'nonius'. Each concentric quadrant is divided into equal parts, whose number is one fewer than on the next outer one. By noting on which of these quadrants the alidade exactly coincides with a graduation, a reading is taken to a fraction of the original division. (Brahe, 1598)

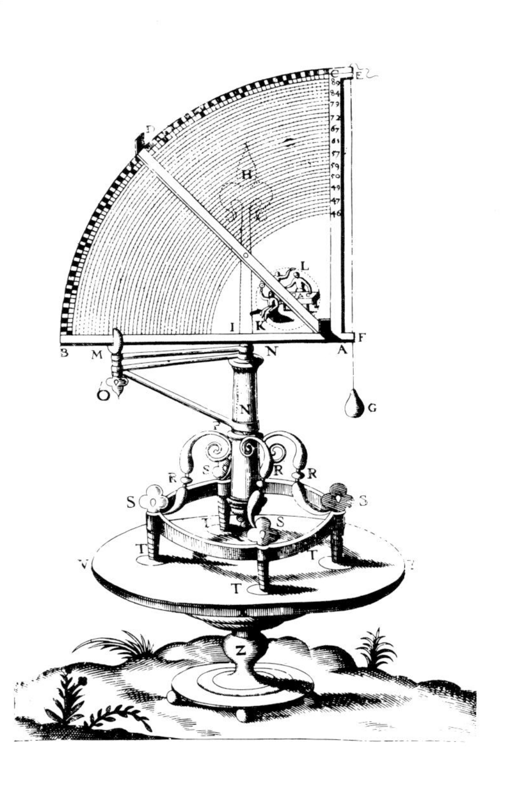

more recent developments, or which showed greater originality on Tycho's part. These included a series of altazimuth quadrants (fig. 13), an altitude sextant (fig. 14) and an altazimuth semicircle, and a sextant (fig. 15) on a large ball-and-socket joint which, like the cross-staff, was not tied to a co-ordinate system, but could take any angular distance. It was in this group that the future direction of astronomical instrumentation was represented, for Tycho's was the last serious use of armillaries or rulers. In subdividing some of the smaller quadrants, he experimented with the method advocated by the Portuguese mathematician and cosmographer Pedro Nuñez (1502–78) (fig. 16), and in one he combined this method of division – the 'nonius' – with transversals.

Modern studies have shown that through a combination of observing techniques, Tycho achieved accuracies down to fractions of a minute of arc. His large instruments operated on the physiological limits of the eye's resolving power, and his achievement represents the culmination of the development of the open-sighted instrument. There would be yet another great observatory where measurements were taken with open-sighted instruments, that of Johannes Hevelius (1611–87) at Danzig. His instruments, however, were the subject of a bitter controversy, which can be understood only in the light of the radical changes in mathematical science and natural philosophy that took place in the century following the foundation of Uraniborg.

2

The Beginnings of Oceanic Navigation

HE IDEA THAT THE MATHEMATICAL sciences formed a coherent domain, with many individual subjects or areas of application related in their common derivation from geometry and arithmetic, was central to the expansion of practical mathematics in the sixteenth century. By the time the movement reached England, this understanding was well developed. The first English translation of Euclid, published in 1570, was prefaced by a long and detailed account by Dee of what the mathematical sciences were, and a demonstration, in an elaborate and expansive scheme, of their relationships and their integrity.

A nevv and neceſſarie
Treatiſe of Nauigation con-
taining all the chiefeſt principles
of that Arte.
Lately collected out of the beſt Mo-
derne writers thereof by M. Blundiuile, and by him
reduced into ſuch a plaine and orderly forme of
teaching as euery man of a meane capacitie
may eaſily vnderſtand the ſame.

They that goe downe to the Sea in ſhips, and occupie their buſines in great waters: Theſe men ſee the workes of the Lord and his wonders in the deepe. Pſalme.107.

17 Title-page to the navigation treatise in Thomas Blundeville's *Exercises*. (Blundeville, 1594)

Dee's preface was an inspirational and influential attempt to preach the unity and importance of the mathematical sciences. The ideas it contains were consonant with the writings of Recorde and the Diggeses, and were the starting-points for a remarkable growth in the publication of practical mathematical works in England in the second half of the sixteenth century, mostly in the fields of navigation and surveying. Notable authors included William Cuningham (1531–86), William Bourne (*fl.*1565–88), Robert Norman (*fl.*1560–96), William Borough (1537–98), John Blagrave (*c.*1558–1612), Thomas Blundeville (*fl.*1560–1602), John Davis (1552–1605), William Barlow (1544–1625) and Thomas Hood (*fl.*1577–96). Instruments, some of them showing genuine originality, were central to these works, the great majority of which were, significantly, published in the vernacular (as, for example, were a number of Fine's books), indicating that the expected audience was practical rather than academic.

The vitality and confidence of this movement suggest a prehistory, during which the relevance of mathematical techniques had been proved in practice, and by far the greater part of this prehistory is found in navigation.

Bearing and distance

Renaissance navigation had two branches – coastal navigation (or pilotage) and oceanic – and the skills involved in each are readily distinguished. Coastal navigation entailed no problems of theory; it was learnt largely from experience of such things as soundings, land-

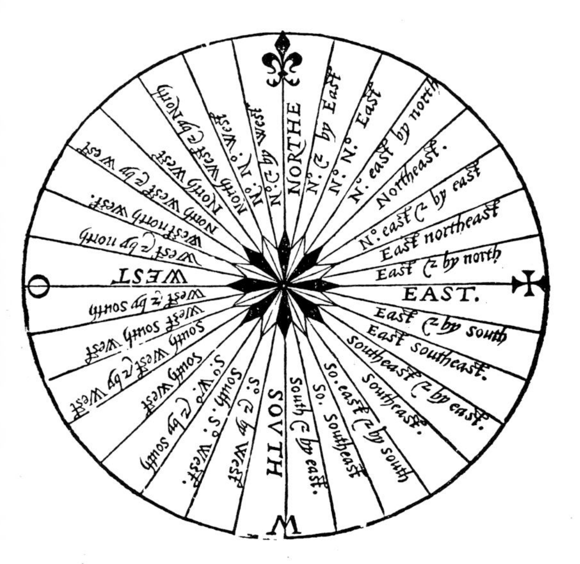

18 The division of the compass according to Martin Cortés into 32 points. Note the fleur-de-lys marking north and the cross east. (Cortés, 1579)

marks, currents and tides. To supplement experience, the pilot had a magnetic compass, a sand-glass to measure the passage of time and to help estimate progress, and a sounding instrument – the common 'lead and line'. He might also have a book of sailing directions or 'rutter' (from the French 'routier'). The compass also had a secondary use, in conjunction with the Sun or Moon, as a rough timepiece, when predicting the state of the tide. Navigation in the Mediterranean followed traditional methods scarcely removed from pilotage.

Oceanic navigation presented quite a different challenge. With no hope of familiar clues to position, the navigator had two options. One was to keep an account of his movements, as best he could, from his last known position – his port of departure or last known sighting. The other was to turn to the heavens and exploit known relations between the measured positions of Sun, Moon or stars and the observer's location. In the absence of a complete solution by either general method, he could try a combination of the two.

Before we consider the beginnings of celestial navigation, we must deal with the primitive instrumentation of pilotage and of 'dead-reckoning' – the traditional accounting from a record of direction (bearing) and distance.

The principal aid to coastal navigation was the lead and line. Written references begin as early as the fifth century BC, with the historian Herodotus, and the cone-shaped sounding lead, first illustrated in the Dutch manual *Spieghel der zeevaerdt* (1584) of L. J. Wagenaer, survived into the twentieth century.

The lead commonly had a hollow indentation in the base, so that 'armed' with a piece of tallow, it would bring up a sample of the sea bed, an important clue for the experienced pilot. Because the Earth's land masses are set on raised platforms in the sea (the continental shelf), soundings give a two-stage forecast of landfall – a distant forecast, the distance varying considerably throughout the world, and a warning of imminent danger. The depth measured was traditionally expressed in 'fathoms', and the fathom was originally the length contained by a man's outstretched arms.

The hand-lead for use in shallow water weighed 7 lb (3.2 kg), a figure that remained fairly constant from at least *c.*1600. Later the 7 lb lead came to be used in boats, while a ship's hand-lead weighed between 10 and 14 lb. These served for soundings of up to 20 fathoms, but for greater depths there was the deep-sea lead, or 'dipsie lead' as the English called it, with a line of 150 fathoms and thinner than the hand-line. The weight used in the seventeenth century was 14 lb (6.4 kg), with heavier weights being tried in the eighteenth. By the later nineteenth century, the deep-sea lead weighed some 28 lb.

Both types of line were being marked by the early seventeenth century. The marking of the hand-line developed into a loose convention of black leather at 2 and 3 fathoms, white cloth at 5 fathoms, red cloth at 7 fathoms, leather again at 10 and white cloth or leather at 15 fathoms. The conventions were modified in the course of time, but complete standardization was never achieved. The deep-sea line was marked with knots at every 10 fathoms.

The division of the circle peculiar to navigation derives, of course, from the instrument fundamental to the technique of dead-reckoning, the magnetic compass, and through the compass to the 'wind roses' on early portolan charts of the Mediterranean. The genesis of the nautical compass is uncertain, but its use in Europe is generally dated to at least the twelfth century, and reference made to a still earlier tradition in China. The story begins with the fabled power of the lodestone, a naturally-occurring magnetic ore, whose very name – lodestone or leading-stone – indicates its directive properties. The earliest nautical application probably involved floating a small lodestone on a piece of wood or cork, but the first surviving accounts speak of an iron needle, similarly floated in a bowl of water, and then magnetized by moving a lodestone, followed by the needle, ever faster round the bowl, before suddenly drawing the stone away. The needle, thus weakly magnetized, would come to rest pointing to the Pole Star or 'Lodestar'.

A later development was to magnetize the needle by stroking it with the lodestone (a process known as 'touching' or 'retouching' the needle) and then to mount it on a pivot. Eventually, the needle was concealed beneath a circular card or 'fly', decorated in a manner similar to the wind roses already used to indicate bearings on charts. By the close of the thirteenth century, the essential features of the nautical magnetic compass were assembled.

While some of the early history is a little speculative, we can certainly say that by the sixteenth century the north-European tradition of a 32–point compass rose, based on the four cardinal points of north, south, east and west, had been commonly adopted. Vestiges survived of the earlier 8–point rose, with the 'rhumbs' or compass bearings named after the appropriate winds of the Mediterranean. (This convention was popular for a longer time in surveying.) North was marked by a fleur-de-lys and east, the direction of the Holy Land, by a cross. A fleur-de-lys marking north and some decoration on the east point survived long after their significance was forgotten. Already the compass was housed in a binnacle – initially simply a wooden box or protective casing – and may have been mounted in gimbals to keep it level. The needle or wire was attached beneath the card, often with some allowance being made for magnetic variation. Since the wire was made of soft iron, a lodestone was an essential accessory to such a compass, and was used to retouch the wire as necessary. Late in the sixteenth century, we find the first reference to the 'lubber's line' – a mark inside the compass bowl, indicating to the helmsman the position of the ship's head.

The discovery of magnetic variation was to have important consequences for navigational theory and practice. It was realized in the fifteenth century that the needle did not point to true north, and compass makers began to allow for this variation by setting needles beneath compass cards displaced by a few degrees. By the early sixteenth century it was clear that magnetic variation was different – both in magnitude and direction – in different parts of the world, and the futility of this practice was realized, though in fact it remained commonplace. The problem of magnetic variation and the conse-

19 (*below*) Seventeenth-century Italian marine compass, in a turned wooden box, diameter 70mm. The needle is beneath the card, which is marked with the initials of the Mediterranean winds. The cross is a 'lubber's line', to be set in the direction of the prow. Greenwich, National Maritime Museum.

20 (*right*) Wooden traverse board, north European, *c.*1800, 305 by 221mm. There are 8 holes for each direction and 4 rows beneath for speeds – 1-12 knots divided to ¼. Greenwich, National Maritime Museum.

21 (*below*) A folding log slate, American, 19th century, comprising two slates, one engraved with columns, in wooden frames, each 255 by 454mm. Cambridge, Whipple Museum.

quences of its study would have important implications for mathematical science and natural philosophy in the sixteenth and seventeenth centuries, and will be considered in some detail in Chapter 4.

The bearings indicated by the compass were recorded by the ship's master. The record might well have been kept on a traverse board (fig. 20), where pegs could be inserted in holes arranged in radiating lines, representing the rhumb lines of the compass. The traverse board was generally made in wood and painted or shaped like a compass rose, with thirty-two rows of holes. Each row had eight holes, one to represent the course steered during each running of the half-hour sand-glass, which was turned eight times during a 4–hour watch.

Traverse boards were introduced in the sixteenth century, and were still being made in the nineteenth century in the Netherlands and in Scandinavia. Some nineteenth-century French boards had pierced brass plates mounted on wood. The log slate (fig.21) was an alternative to the traverse board for keeping such records, and was more popular in America. The slates were mounted in wooden frames – often two slates were hinged together – and were sometimes engraved with headed columns, commonly: 'H[our], K[nots], F[athoms], Courses, Winds, L[ee] W[ay]'. Leeway was the allowance made for drift to leeward.

The watch-glass was generally arranged to run for half an hour, but 1–hour and 4–hour glasses are also found. They can be set in ivory, leather, metal or, most commonly, wooden stands, but are rarely signed or dated. The joint between the bulbs can give some indication of date. Until the eighteenth century, a diaphragm was placed between the bulbs, and the joint sealed with wax and bound with canvas and thread. Around the mid-eighteenth century, the technique was introduced of fusing two separately made bulbs, with a pierced brass bead inserted to regulate the flow. It was only later in the century that a glass could be blown in one operation.

At the end of the watch, the master would calculate the distance travelled, from estimates of the ship's speed, and determine the course 'made good'. The traverse board would then be cleared for the next watch. One of the first practical achievements of English navigation, however, was the introduction of an instrument which allowed the traditional estimate to be replaced by a measurement of speed.

The 'log' was first described in the earliest truly English manual of navigation, William Bourne's *Regiment for the sea* of 1574. The log or 'log-ship' was a piece of wood, usually in the shape of a quadrant (some were carved to the shape of a fish) and weighted on its curved side to make it float upright. It was attached by three short pieces of rope to a long rope, called the log-line. This line, soon to be supplied with knots at regular intervals, was wound on to a hand-held reel. In use, the log was cast astern and allowed to clear the turbulent water immediately behind the ship, before the seaman counted the number of knots played out during half a minute, timed by a sand-glass (fig. 22). A refinement was to secure one of the attaching ropes to a plug set in the log, so that a sharp tug would pull it out and the log, lying flat in the water, could be hauled in more easily.

22 Two log glasses, in wood (135mm) and brass (82mm) mounts, both marked '28', to indicate the running time in seconds. Cambridge, Whipple Museum.

The line was supposed to be knotted at intervals such that the number of knots thus noted was a direct indication of speed in miles per hour, where the nautical mile was the length of one minute of latitude measured along a meridian on the Earth's surface. Seamen traditionally knotted the log-line every seven fathoms, but since this figure depended in theory on the opinions of astronomers and surveyors, there was much subsequent discussion of the correct interval between the knots. The use of the log is, of course, the origin of expressing a ship's speed in 'knots'.

An alternative 'instrument' was the 'Dutchman's log', popular among seamen from the Netherlands during the seventeenth and eighteenth centuries. A float cast overboard in the bows was timed by a sand-glass, the pulse-beat or a repeated incantation, over a marked distance on the ship. Conversion tables from times to speeds are found engraved on such everyday objects as tobacco tins (fig. 23).

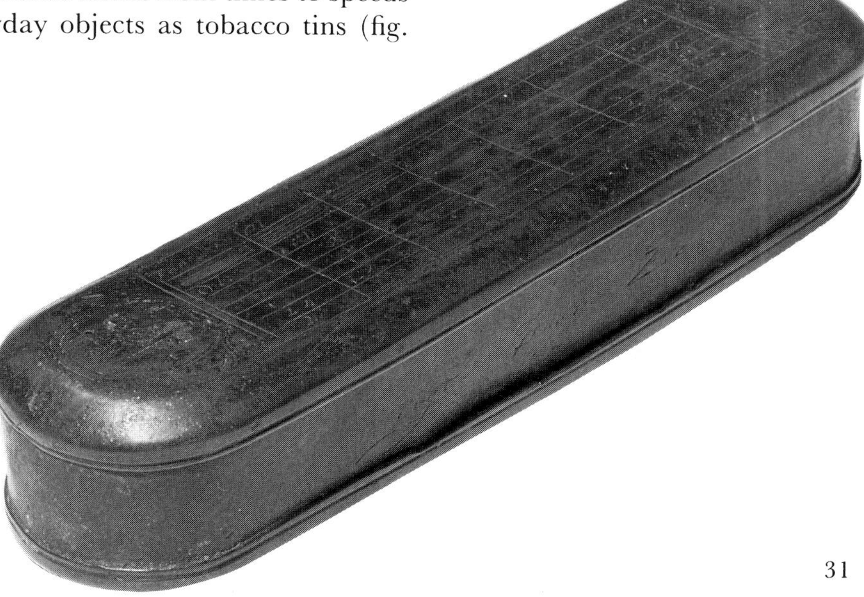

23 The 'Dutchman's log' in the form of a tobacco-box, length 175mm. As well as a perpetual almanac and lunar calendar, it carries a table for calculating a ship's speed from the time taken for a float to pass a measured distance.

The log-glass was at first recommended to run for half a minute, but many surviving examples are slightly faster. Revisions of the accepted estimate of the size of the Earth, and thus of the length of a degree, meant that from the mid-seventeenth century, mathematicians were advocating a greater distance between the knots. The seaman's natural convervatism, however, was such that the traditional distance of seven fathoms remained, and the only practical solution was to shorten the duration of the log-glass. A glass marked '28', for example, indicates the measured time in seconds.

With the coming of the log, traverse boards were made with extra holes, beneath the rose, to record the speeds measured whenever the log was cast during the watch. There was still much uncertainty: eddies created by the ship's motion would affect the log, the art of playing out the line was not trivial, some allowance had to be made for the effects of currents, and so on. But the figure for distance travelled, computed by the master when he cleared the traverse board for the next watch, now had some basis in measurement.

Such were the aids to traditional navigation based on bearing and distance. They served moderately well on established routes over charted seas, but were ill-suited to exploration. Portuguese navigators of the fifteenth century, their exploratory voyages extending ever farther down the west coast of Africa, had been obliged to pioneer new ways of determining positions and of returning to them in the future.

The 'new' navigation and the problem of latitude

The ambitions of the Portuguese demanded new navigational techniques, and through the initiative and patronage of Prince Henry the Navigator (1394–1460), these were deliberately sought in astronomy. Astronomers had long been familiar with the straightforward relationships between the appearance of the heavens and the latitude of the observer – we have seen these built into such portable instruments as the torquetum and the astrolabe. Thus, they had a ready theoretical solution to at least one of the unknowns of exploratory navigation, and in its application we find a good deal of familiar instrumentation adapted to new ends.

Of the European nations, it was thus the Portuguese who first forged a useful working relation between astronomy and navigation, and who institutionalized it in such posts as those held by Pedro Nuñez, a royal cosmographer and university professor of mathematics. They were followed by the Spanish, whose need arose from their trans-Atlantic enterprises of the sixteenth century. They too established official positions for cosmographers as well as relevant university professorships. Cosmography was a useful catch-all mathematical discipline, comprising geography, astronomy, surveying and navigation.

Seamen of other European nations first learnt the methods of the new navigation by reading translations of Spanish textbooks. Most famous among the Spanish theorists were Pedro de Medina (1493–1576) and Martin Cortés (d.1592). Medina's *Arte de navegar*, first published in 1545, was translated into German, French, English and Italian. Cortés' *Breve compendio de la esfera y de la arte de navegar* appeared in 1551, and there were nine English editions between 1561 and 1630.

To the astronomer the most obvious measure of a terrestrial latitude was the altitude of the Pole Star (Polaris), for which a portable version of the astronomer's quadrant was at first recommended. This was a quarter circle in wood or brass, graduated in degrees along the limb, with two pin-hole sights on one straight edge and a plumb-line attached to the apex. Altitudes were taken by sighting through the pin-holes, clamping the plumb-line with a finger, and noting its position on the degree scale.

Polaris does not, however, mark the celestial pole precisely, but makes a small daily revolution around it (fig. 24); in 1500 it was some 3½° distant. An early expedient was to use the quadrant, not to measure latitude as such, but to indicate the height of the pole for certain ports, or to measure relative distance north or south. The navigator had to mark the limb with the plumb-line positions of ports, and to remember to use his quadrant only when the two circumpolar stars known as the Guards or Pointers were in a particular configuration with respect to Polaris. This might be sufficient for his needs; otherwise the degree scale, with a simple conversion rule, allowed him to determine distances in leagues – much more familiar than latitude difference in degrees.

Before long a rule was evolved for more general use – the so-called 'Regiment of the North Star' – to correct for the displacement of Polaris from the celestial pole. The 'Regiment' encapsulated, in an easily remembered form, the number of degrees to be added or subtracted from the measured altitude of Polaris for eight estimated positions of the Guards in their daily revolution.

There was a more immediate, practical problem: the quadrant proved all but impossible to use at sea, and the navigator had to go ashore to make any worthwhile observations. The introduction of the mariner's astrolabe (fig. 25), late in the fifteenth century, was an improvement. Like the quadrant, it was a primitive adaptation of an astronomical instrument – the planispheric astrolabe, stripped of all sophistication and left with only two features of the verso, the degree scale and the alidade. The astrolabe was better suited to shipboard use than the quadrant, but must still have been difficult to use in all but the calmest conditions. Suspended from the thumb by a shackle, it was often weighted at the bottom, and had all extraneous metal cut away to decrease air resistance. By the mid-sixteenth century the standard instrument was of cast brass.

The astrolabe had a further advantage over the quadrant: it was much more convenient for taking the altitude of the Sun. As the Portuguese voyages extended farther down the African coast, Polaris approached the often hazy horizon and became less useful for fixing latitude. Astronomers were obliged to devise an alternative. In theory any visible heavenly body would have sufficed, provided it was covered by an adequate predictive theory, and provided its altitude could be measured at a known point in its daily apparent motion.

24 The use of the cross-staff by a mariner to determine the altitude of the Pole Star. The altitude of the pole is a direct measure of the observer's latitude, but note that the Pole Star is shown describing a circle daily around the true pole, so that a correction must be applied to the measured altitude. (Medina, 1563)

25 A mariner's astrolabe by Elias Allen, 1616 The openwork frame is weighted at the bottom, and the scale is divided for direct measurement of altitude, that is from 90° at the top to 0 to 90 to 0 to 90 again. Subdivision is to ¼° and one quadrant is divided by transversals to 10 minutes, which represents an unrealistic level of accuracy. St Andrew's, University, Department of Physics.

The most straightforward point to choose was where the body crossed the observer's meridian, identified when it reached its maximum or minimum altitude, and the Sun would, of course, be particularly useful for daytime observation.

We have already met with solar declination in Chapter 1 – the plane of the ecliptic is inclined to the equator, from which the Sun 'declines', i.e. it undergoes an apparent annual cycle of movement away from the earth's equatorial plane. For the sixteenth-century navigator, it was a considerable challenge to be faced with tables of declinations for every day of the year, giving the correction which had to be added or subtracted (according to the date and to whether the ship was north or south of the equator) to the observed meridian altitude. The altitude was taken with the astrolabe: the spot of light from the upper pin-hole on the alidade was arranged to fall directly on the lower one and, observing at about noon, the maximum reading was noted from continuous observation. Latitude was the zenith distance of the equatorial plane, obtained from the adjusted altitude measurement.

Whatever difficulties the seaman may have experienced in learning to find latitude from the Sun, by 1500 a general solution to the latitude problem had been found, and further developments would improve the tables and instrumentation. Latitude could be discovered either from the altitude of Polaris, applying a correction for its distance from the pole (the Regiment of the North Star), or from the zenith distance of the Sun, found by observing its altitude at noon and applying the declination found in the calculated tables (the so-called 'Regiment of the Sun').

A further direct contribution from the astronomers was the adaptation of the cross-staff (figs 24, 26) to navigational use. The cross-staff was readily manufactured, light to use, easily stowed, applicable to different types of observations, and it became the seaman's preferred instrument for taking altitudes. It could be used either for Polaris or for the Sun; in the latter case ('shooting the Sun', as the observation was called) the navigator would sight the upper edge of the Sun's disc and apply a correction for its semi-diameter. He could protect his eyes by fixing a piece of smoked glass to the upper end of the cross.

So far we have seen the direct application of astronomical techniques. Instruments – quadrant, astrolabe and cross-staff – were simplified and handed down by the astronomers, together with the necessary rules and tables. These techniques and their instrumentation were introduced to a general English audience through Richard Eden's translation of Cortés' textbook, and it is interesting that the first significant contribution by the navigator himself to latitude-finding came from the relatively new English school. Like its earlier contribution, the log, it was a practical device by a practical man.

There were difficulties and inconveniences in the use of the cross-staff. In theory, it required the navigator to sight star or Sun together with the horizon, which was, of course, a difficult skill. The apex of the angle to be measured was within the eye; poising the staff on the cheekbone was therefore a critical, but always inexact, technique. Looking directly at the Sun was uncomfortable, and it was also difficult to maintain the instrument in a vertical plane while keeping a steady posture on deck. For high altitudes it was very inaccurate.

In 1585 Captain John Davis described in his book *The seaman's secrets*, two designs for an instrument (fig. 27) for measuring solar altitude using the Sun's shadow rather than a direct sight. Since the navigator, with his back to the Sun, sighted the far horizon, the instrument, though sometimes referred to as the 'Davis quadrant', became generally known as the 'backstaff'. It settled down into a standard pattern (fig. 28) of two main 'limbs', two cross braces and two arcs. The limbs are often in ebony, sometimes rosewood, pearwood or mahogany, the arcs, and occasionally the braces, in boxwood.

The larger arc (usually calibrated to 65°, but commonly referred to as the '60–degree arc') carried the movable shadow vane, while the smaller (calibrated to 25°, but known as the '30–degree arc') had a pin-hole sight. Attached to the furthest end of one limb was the horizon vane with a horizontal slit-sight. To measure the Sun's meridian altitude the seaman first set the shadow vane to a division on its arc (generally divided to degrees) some 15 to 20° less than the

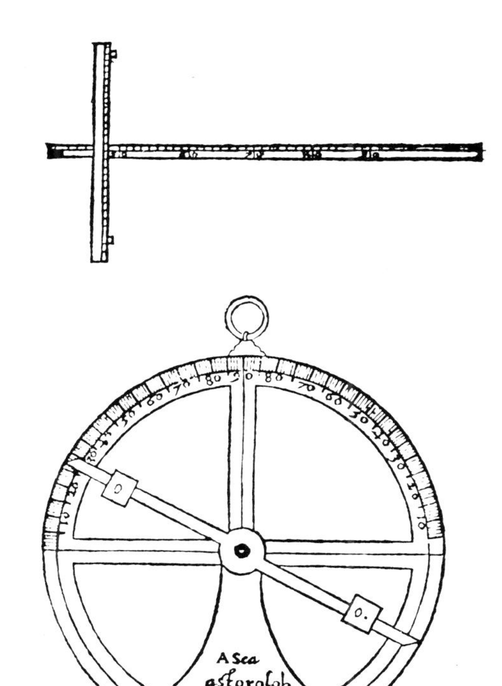

26 The cross-staff and mariner's astrolabe illustrated in William Bourne's *Regiment for the sea.* (Bourne, 1574)

27 A form of backstaff, illustrated in John Davis, *The seamans secrets.* While the shadow of the end of the upper index is made to fall on the horizon vane, the lower end of the lower is brought into coincidence with the horizon vane and the horizon. Note that, as with the cross-staff, index pieces slide along the staff, which is divided. (Davis, 1595)

28 Backstaff by E. Blow of London, 'in Plow Alley at Vuion Staires Wapping', 1736, length 545mm. The limbs are rosewood, the arcs, braces and vanes box. The 30-degree arc is divided to 5 minutes and by transversals to 1 minute; there is a solar declination table on the reverse. It is unusual for the original vanes to have survived, but here there is also the bonus of a rare 'Flamsteed glass'. An owner's plate records that it was sold the year following manufacture to a Mr Thomas Halcott. Cambridge, Whipple Museum.

expected altitude. With his back to the Sun, and his instrument held vertically, he allowed the shadow cast by the shadow vane to fall precisely on the slit of the horizon vane. At the same time, he moved the sight vane on the 30–degree arc so as to see the horizon through its pin-hole and the slit in the horizon vane. The observation continued as the Sun moved across his meridian, and the meridian altitude was the sum of the maximum reading on the lesser arc and the set reading on the greater.

It was common practice to provide a transversal scale on the 30–degree arc, with a basic division to one degree, subdivided to 5 minutes, and by transversals to one minute. In practice this order of accuracy was quite unrealistic. It was usually calibrated for zenith distance (65 – 90°) as well as for altitude, and may also have a table of solar declination on the reverse. The 60–degree arc sometimes has divisions at 5–degree intervals on its rim, which are slightly offset from the main degree scale. This allows for the fact that the shadow will register the upper edge of the Sun's disc and not the centre. A later improvement, usually ascribed to the British Astronomer Royal John Flamsteed (1646–1719), was to substitute for the plane shadow vane, one containing a convex lens (called a 'Flamsteed glass'), which would focus a spot of light onto the horizon vane. This was recommended for use in hazy weather.

Latitude sailing

The application of methods derived from astronomy had effectively solved the latitude problem. The instruments, techniques and people involved, have underlined the intimate relation between astronomy and navigation, just as they demonstrated in their day the integrity of the mathematical sciences. The solution was adequate to the navigational needs of the African explorations of the Portuguese, sailing down a roughly north-south coastline. More ambitious expeditions into the Atlantic would of course raise the far more intractable problem of longitude.

With the heavens united in a daily apparent rotation, there was no reference point for fixing east-west positions, corresponding to the pole for the north-south co-ordinate. The problem of longitude was neatly explained in Bourne's *Regiment*:

> Nowe some there be that be very inquisitiue to haue a way to get the longitude, but that is to tedious. For this they must consider, that the whole frame of the firmament is carried round from the east to the west in .24. hours, so as ther remaineth neither light not marke, but goeth rounde, sauing only the .2. poles of the world, and theses .2. stand alwayes fast. (Bourne, 1963, p.238)

The general plan of attack that had solved the latitude problem was tried with longitude: methods sound in theory are found proposed by such men as Apianus, Werner, Gemma Frisius, and so on. While they had no practical success, their pursuit, as we shall see in Chapter 4, would have important implications for science.

Consider for the present the instrumental aids available in practice to the navigator at the end of the sixteenth century. He could discover, in spite of cloud, the ship's direction from his steering compass. With a log, he could measure her speed, and record both direction and speed on his traverse board. After making what allowance he could for various disturbing influences, he was able to plot his estimated position on a chart. If conditions permitted, he would first have checked the course made good by a latitude sight, using his preferred altitude-measuring instrument on the Sun or the Pole Star. The result of the dead-reckoning may well not have accorded with the observed latitude, but the former would have to accommodate the latter, since, within the limits of instrumental error, it was the latitude that could be directly measured.

Given the available instruments and their relative degrees of accuracy, a technique of navigation, known as 'latitude sailing' or 'running down the latitude', naturally emerged. It may be regarded as the technique that followed that of bearing and distance. The navigator set a course to bring his ship to the latitude of his destination, but well to either the east or the west. He had a rule to tell him the distances to sail in different directions in order to increase or decrease his latitude by one degree. Having gained and checked his latitude, he then maintained it while sailing east or west until he sighted his landfall. This practice is a compelling illustration of the mariner's ignorance of his longitude, and it survived, with only marginal refinement, until the longitude was found.

3

The Impact of Geometry on Surveying

SURVEYING WAS THE LAST OF OUR THREE mathematical sciences to be changed in character and method by the early modern revival of mathematics, and it changed with some reluctance. Its established practices operated at a relatively primitive level, but were adequate to its needs. Surveys were undertaken for practical reasons, connected with road-building, estate evaluation, rent-fixing, boundary control, and so on. The surveyor was concerned, not only with land measurement, but with estate estimates and valuations of all kinds. Renaissance mathematicians, however, saw in surveying an opportunity to demonstrate again the practical usefulness of geometry, and the importance of mathematical science. Success would require a propaganda effort directed at the surveyor – who, as likely as not, preferred to be left in peace with his familiar techniques – and at his patron. They would need to be convinced of the value of a new type of surveying, requiring mathematical skills, and of a new image of the surveyor as a geometer, whose badge of office would not be a notebook and pole, but a theodolite.

Ancient and medieval survey

Evidence for the methods of ancient surveying – Egyptian, Greek and Roman – comes principally from accounts of instruments, rather than of surveys, and it is difficult to be sure to what extent they represent general practice. Even so, the instruments concerned are, for the most part, fairly primitive. The 'groma', used by the Egyptians and Romans, consisted of a right-angled wooden cross, suspended and resting horizontally, with plumb-lines hanging from the four extremities. This is an early form of the 'surveyor's cross', which, in a variety of forms, became a standard instrument, used to establish a direct line of sight, as well as subsidiary lines or 'off-sets' at right angles.

Vitruvius, the Roman architect of the first century BC, describes a level consisting of a long plank of wood, adjusted by means of plumb-bobs, which are made to coincide with pre-drawn perpendicular lines. Alternatively, a groove in the plank could be partly filled with water, which should rest parallel to the top. The most famous instrument described by Vitruvius was his 'hodometer' – a surveying wheel used to measure distance along the ground, each revolution being recorded by a stone released into a box. Instruments working on the same general principle are used today, but down the centuries they have been known by a good many different names – surveyor's wheel, hodometer, odometer, perambulator and way-wiser.

The most sophisticated surveying instruments recorded in ancient times were the dioptra and water-level described by Hero of Alexandria (*fl.* AD 62). The former was a divided circle with a centrally pivoted alidade, generally used in a horizontal position, but supplied with two rack-and-pinion motions, in altitude and azimuth. The level had a horizontal tube in a long wooden mount, connected to upright glass tubes at both ends. Brass sights with vertical screw adjustments could be made to coincide with water levels in the tubes. These instruments indicate a remarkable potential for sophistication but there is no evidence that they had much effect on surveying practice.

Medieval survey was largely a matter of direct measurement with poles and ropes, perhaps supplemented by primitive instruments of the Roman type. Texts show instruments derived from astronomy being used for measuring the heights of towers, and so on, but these illustrated rather the power of geometry than the practice of surveyors. It was a distinction that would trouble the mathematicians for some time to come.

The early influence of geometry

Some astronomical instruments could be applied directly to surveying. The reverse of the astrolabe carried both a full circle and a shadow square, either of which could be used with the centrally pivoted alidade. Treatises on the instrument frequently explain and illustrate its uses in surveying – for measuring the heights of towers or the depths of wells, or for taking bearings.

In order to take bearings or azimuths, the astrolabe would be mounted in a horizontal position, and angles indicated by the alidade were read off the degree scale. In 1533, in a tract bound up with the second Gemma Frisius edition of Apianus' *Cosmographia*, Gemma described the principle of triangulation, whereby bearings taken on

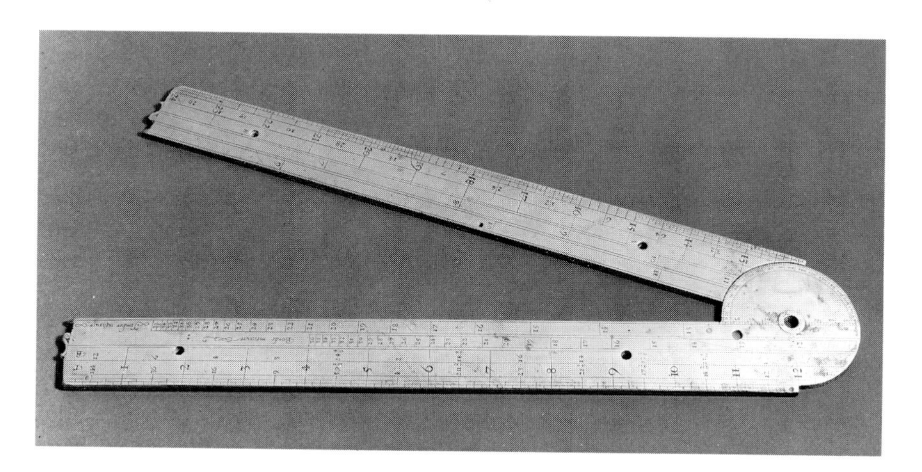

29 A surveyor's rule by Humphrey Cole, 1574, folded length 338mm. A general purpose instrument with a number of scales that relate to the measurement of land and timber. The 180-degree scale allows azimuth angles to be measured when pairs of sights are fitted into the holes in the arms. The arms can be set at 90°, when altitudes can be taken by a plumb-line and quadrant on the reverse side, or set at 180° and used as a plane table alidade. The scales follow those recommended in Digges, 1556. Cambridge, Whipple Museum.

each of a series of targets from two vantage points will locate them all in relation to each other and to the two stations. A single measured or known distance will give a scale to the map thus drawn. Alternatively we might say that bearings taken from either end of a measured base-line can be used to locate an object at an unmeasured (or unmeasurable) distance. By a series of such triangles an entire map can be constructed without further linear measurement.

30 Compass by Jacopo Lusverg, 1687, 120 by 120mm, divided to 1° and marked with the winds of the Mediterranean.

To maintain a consistent orientation for the astrolabe when taking the bearings required for triangulation, a compass needle had to be attached to or inserted into the instrument, and Gemma's nephew Gualterus Arsenius (*fl.*1555–75) was one of the first instrument-makers to produce astrolabes with inset compasses. Clearly a simplified instrument, specially adapted to surveying, would be more convenient to use and cheaper to make. The adaptation and simplification of the astrolabe for surveying is analogous to the emergence of the mariner's astrolabe in navigation.

Gemma described a wooden instrument with a graduated circle and a centrally pivoted alidade. In use, a detached compass was rested on it. A compass was incorporated into a similar instrument described by the Italian mathematical scientist Niccolò Tartaglia (*c.*1500–57) in his *Quesiti et inventioni diverse*, published in 1546 but written some time before. In this case, the circle is not divided by degrees (though his text indicates this possibility), but by a division based on the traditional winds of the Mediterranean (fig. 30). His cardinal directions are thus marked: 'T[ramontana], G[reco], L[evante], S[irocco], O[stro], G[arbino], P[onente], M[aistro]'. Each of these eight sections is divided into nine parts, giving a total division of the circle into seventy-two parts.

William Cuningham, in his *Cosmographical glasse* of 1559, explained the principle of triangulation, with due acknowledgement to Gemma, and described an instrument, which he called the 'geographicall plaine sphere'. This was a circular plate, 'made muche like the backe parte of an Astrolabe' (Cuningham, 1559, p.136), with a centrally pivoted alidade. The circumference was divided into 360° (arranged in quadrants running 0 to 90 to 0 to 90 to 0), but the whole face was also engraved according to the north-European nautical division (cf. Tartaglia) of thirty-two points. There was also a small inset compass for orientation. With Cuningham's instrument, the rationalization of the astrolabe into a surveying instrument was complete.

Azimuth instruments of this general type (fig. 33), which by the seventeenth century were fairly well accepted among surveyors, became known as 'theodolites'. The term was adopted by Digges in his *Pantometria* of 1571, where such a circle, with an inscribed double shadow square (a term which is explained below), is called the 'theodelitus'. This has caused considerable confusion in modern times, since a theodolite is now regarded as an altazimuth instrument. In reading the primary literature, however, it is important to remember that any unqualified reference to a 'theodolite' made at least until the end of the eighteenth century almost always indicates an instrument with an azimuth circle only.

I will use the term 'simple theodolite' to indicate unambiguously

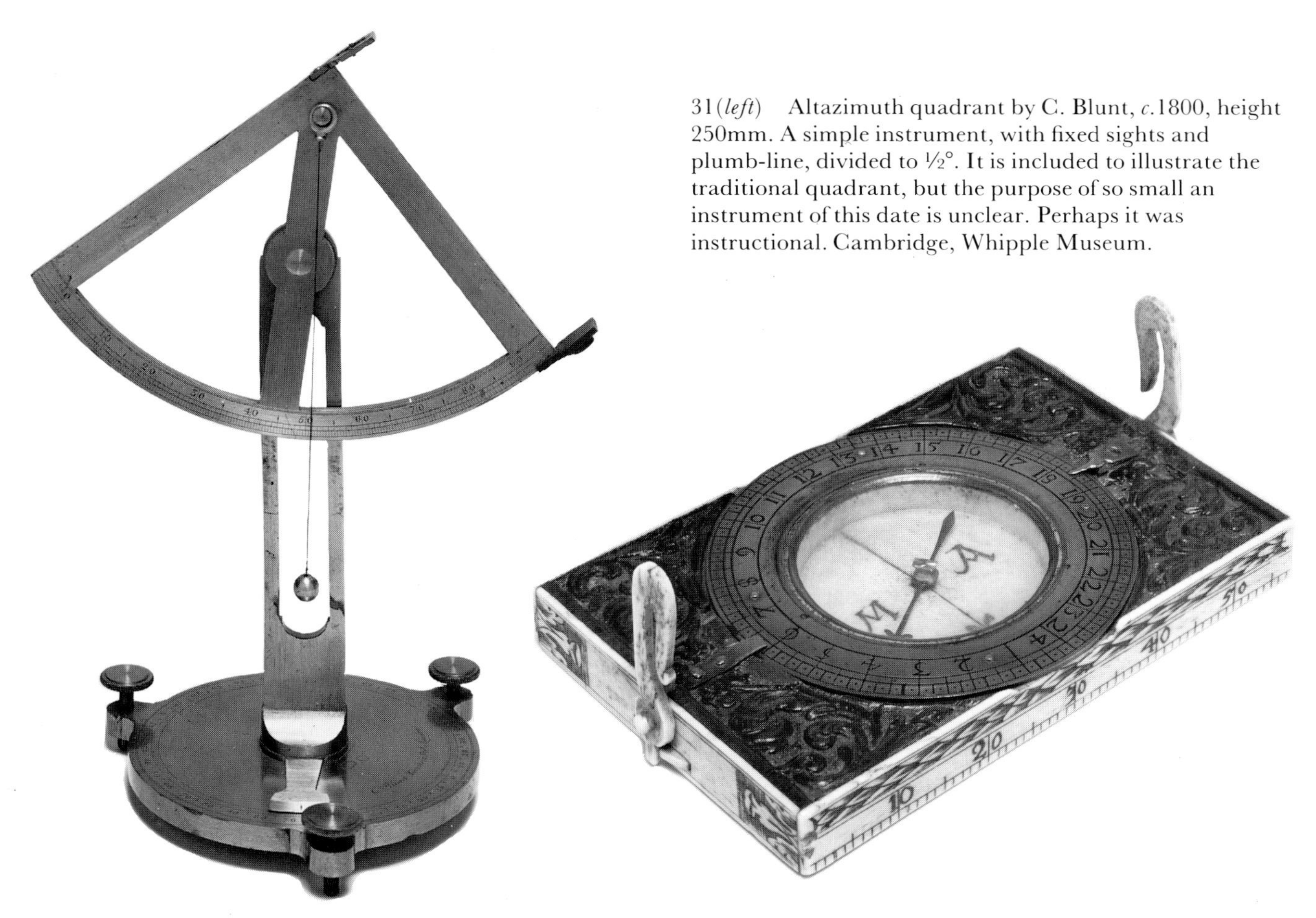

31 (*left*) Altazimuth quadrant by C. Blunt, *c.*1800, height 250mm. A simple instrument, with fixed sights and plumb-line, divided to ½°. It is included to illustrate the traditional quadrant, but the purpose of so small an instrument of this date is unclear. Perhaps it was instructional. Cambridge, Whipple Museum.

32 (*right*) German miner's compass, 17th century, illustrating on the brass ring the traditional division of the circle into 24 parts, described in the mining textbook of Agricola (Agricola, 1556). The two halves of the compass are marked 'M[orgen]' and 'A[bend]' and the brass ring can be turned according to the compass direction. The ivory hooks serve as sights or for attaching the compass to a line. Cambridge, Whipple Museum.

an azimuth instrument, where bearings are indicated by the positions of an alidade or sighting rule on a divided circle. This term was used by makers such as Stanley and Watts in the late nineteenth and twentieth centuries, as by then 'theodolite' had taken on its modern sense. The modern use in the secondary literature of the term 'circumferentor' to refer to this class of instrument has been particularly unfortunate, because the word was used consistently from the seventeenth to the twentieth century, to indicate a quite different instrument. Further, the distinction between '[simple] theodolite' and 'circumferentor' was, as we shall see, an important one for the practice of surveying and the development of its instrumentation, and is worth preserving in order to understand its history.

First, however, we must look at further direct applications of astronomical instruments. A portable quadrant (fig. 31), similar to the type adapted to navigation, could be recommended also to surveyors, since it might be used readily to measure altitudes and, with

33 A simple theodolite by Danfrie, *c.*1600, diameter 194mm. There are eight primary divisions, each one marked with the name of a wind and divided further from 0 to 45°. The small compass in the centre is for orientation. Sights of this type are called 'slit and window sights'; a vertical wire is missing from the fore-sight. Cambridge, Whipple Museum.

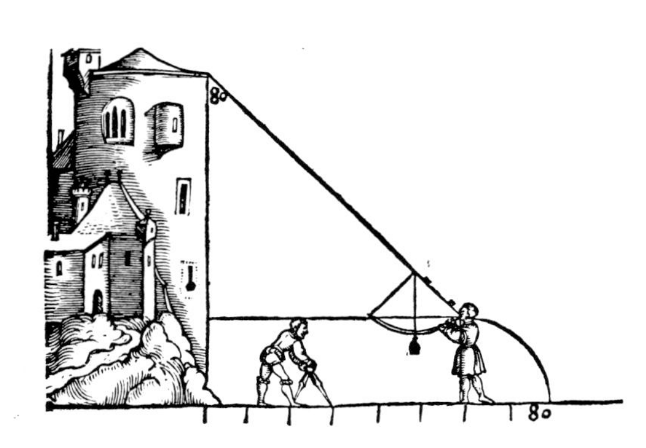

34 (*left*) The quadrant used to measure the height of a tower. In the simple case, where the top is sighted through the fixed pin-holes and the plumb-line bisects the limb, the height of the tower equals the distance of the observer from its base. Note that a correction is made for parallax: the observer's eye height is added to the horizontal distance. (Apianus, 1532)

35 (*right*) A more complex case than in fig. 34; the base of the tower is inaccessible, and its height is determined from the angle it subtends at two stations a measured distance apart. Note that the quadrant readings (50 and 25) are not in degrees, but are taken from the shadow square ('umbra versa'). (Apianus, 1532)

a little ingenuity, to take bearings. A shadow square was generally included if surveying was considered a possible application; the square might be inscribed within the arc of the quadrant, or the arc inscribed within the square.

The shadow square itself, whether inscribing or circumscribing the quadrant arc, comprised a square with two adjacent sides divided into equal parts (figs 34, 35, 36), and was used to measure ratios (which we might regard as tangents or co-tangents, expressed as fractions) rather than angles in degrees. The common corner of the undivided sides either carried the pivot of an alidade or the suspension point of a plumb-bob; in the latter case, fixed sights would be fitted to one edge of the instrument. The former type, having an alidade, was commonly used with squares circumscribing quadrant arcs; they had to be 'levelled' by a plumb-bob, but the readings were indicated by the alidade. With the latter type, readings were indicated by the plumb-line itself.

On early instruments there are usually twelve divisions on each side, marked by lines drawn as if radiating from the pivot or suspension point. The scales each run from zero to a common 12 at their intersection, and they are given different names to distinguish their possibly ambiguous readings. The scale measuring ratios (tangents) from 0/12 to 12/12 (angles from 0 to 45°) is the 'umbra versa' (back or contrary shadow); that measuring ratios from 12/12 to 12/0 (angles from 45 to 90°) the 'umbra recta' (direct or right shadow).

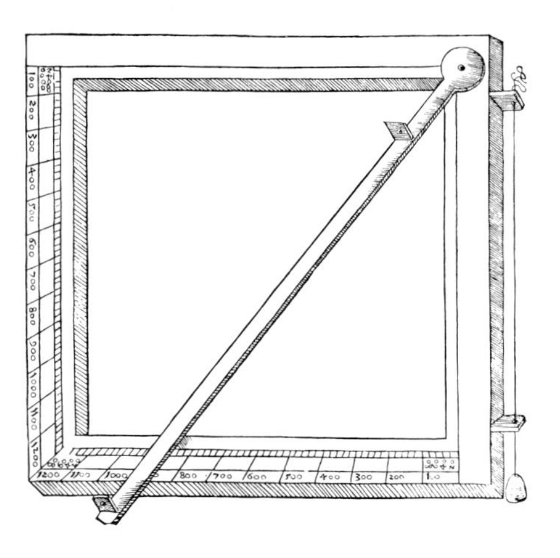

36 Peurbach's geometrical quadrant, with pivoted alidade and plumb-line. Angles are measured in terms of ratios or tangents. (Regiomontanus, 1544)

Alternative names were used for such an instrument – 'geometrical square' or 'geometrical quadrant' were fairly common – but 'shadow square' derives from one application to surveying, which involved measuring the shadow cast by a perpendicular structure, such as a tower, and using the instrument to determine the ratio between this and the height. A sight is taken on the Sun by targeting the point of light from the fore-sight on to the near-sight. In the simplest case, where the alidade or plumb-line then bisects the square (i.e. the solar altitude is 45°) the height of the tower equals the length of its shadow: their ratio is given by the instrument as 12/12. With the Sun at elevations less than 45°, the ratio of shadow

to height is given by the umbra versa scale, with 12 as the denominator; the umbra recta scale applies for angles greater than 45° and 12 is the numerator of the fractional ratio.

The shadow can, of course, be neglected, the top of the tower sighted directly and a linear measurement taken from the observer's station. In this case, an allowance must be made for the height at which the instrument is held. The use of the shadow eliminates this source of error.

It is not clear to what extent the shadow square was used in practice, but it is an ubiquitous feature of early mathematical instruments. Its association with surveying is a subtle indicator of technical competence among surveyors. They were familiar enough with ratios and the rules for applying them: such a degree of technical competence was required for drawing plans to scale, and they would shortly have a calculating instrument – the sector (see Chapter 4) –

37 (*left*) Sinical quadrant by Domenico Lusverg, 1717, radius 158mm, divided to ½° and read by a vernier. There is also the equivalent of a shadow square, divided 0–100–0, and the network of lines in the centre is used with the divided arms for trigonometrical calculations. There is a ball-and-socket mount for a staff or tripod. Cambridge, Whipple Museum.

38 (*right*) The reverse of the quadrant in fig. 37 has a gunner's quadrant, used with a plumb-bob, and a diagram to show how it can assist in setting up a cannon. Cambridge, Whipple Museum.

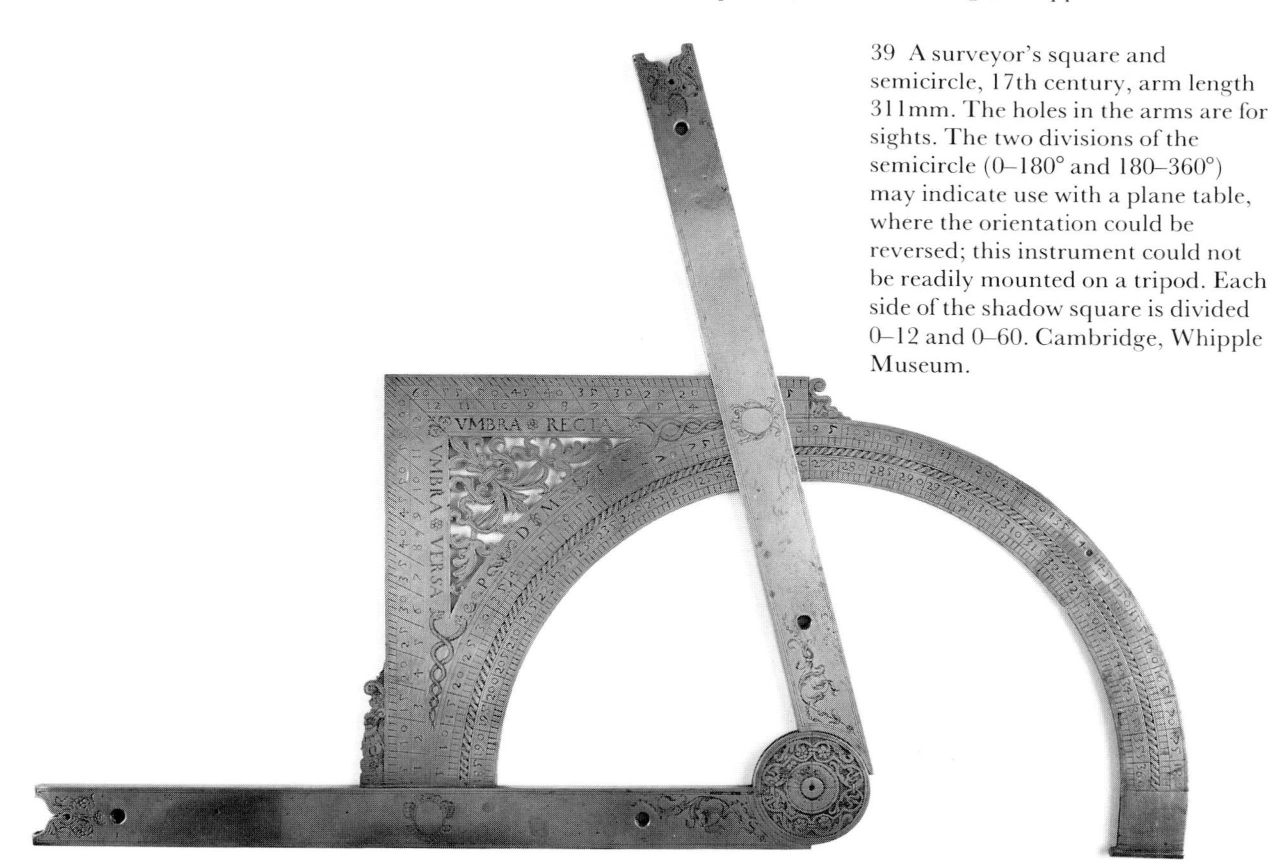

39 A surveyor's square and semicircle, 17th century, arm length 311mm. The holes in the arms are for sights. The two divisions of the semicircle (0–180° and 180–360°) may indicate use with a plane table, where the orientation could be reversed; this instrument could not be readily mounted on a tripod. Each side of the shadow square is divided 0–12 and 0–60. Cambridge, Whipple Museum.

specifically designed for handling ratios. They were much less comfortable with angles and the mysteries of trigonometry.

A third instrument, the cross-staff, completes the trio of astronomical instruments – astrolabe, quadrant and cross-staff – which were applied, with modifications, to surveying as well as to navigation (fig. 53). A number of the mathematicians we have encountered promoted cross-staves modified for surveying use. Apianus described one in 1533; Gemma's staff was designed as a universal instrument with applications in surveying; Leonard Digges included a 'profitable staffe' among the instruments recommended in his *Tectonicon* of 1556; Fine's surveying treatise *De re & praxi geometrica* of 1586 describes a geometrical quadrant – a quadrant with a circumscribed shadow square and alidade – and a 'baculus mensorius' or cross-staff.

Thus the early attempts to bring surveying within the domain of practical geometry, and to demonstrate its status as a mathematical science, involved establishing a congruence between its instrumentation and that of astronomy. Simplification could render these instruments genuinely useful to the surveyor, if a little bewildering. But the attempt to 'geometrize' surveying practice was taken further, with the design and promotion of exotic instruments, incompatible with the needs and capacity of the ordinary surveyor.

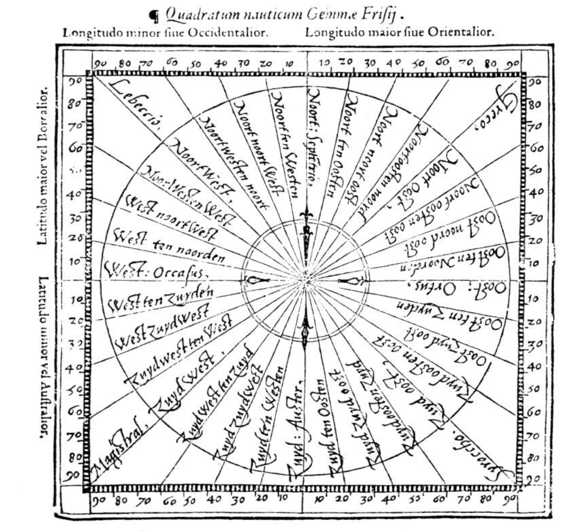

40 The nautical quadrant of Gemma Frisius, where the mariner's division of the circle is shown in relation to the shadow square, to form an instrument for use in course and distance calculations. The value of such an instrument was limited by its assumption of a flat Earth – a limitation appreciated, of course, by its inventor and users. (Apianus, 1550)

Exotic instruments

An early instance of ambitious instrumentation, cited uncritically in modern secondary works, is the altazimuth theodolite – an instrument designed to take altitudes and azimuths at the same time. The earliest altazimuth instrument specifically intended for surveying seems to have been the 'polimeter' of the German cosmographer Martin Waldseemuller (1470–1518), described in the 1512 edition of the encyclopaedic textbook *Margarita philosophica*. Here a sighting tube, in the form of a triangular prism, is mounted on a vertical semicircle, moving in altitude, with an inscribed double shadow square and plumb-bob. The whole pivots above a horizontal circle, with four inscribed shadow squares, themselves comprising a square. By a single sighting, altitude and azimuth could be taken together, either as degrees or as fractional ratios.

A similar instrument is described in Leonard Digges's *Pantometria* of 1571 and named the 'topographicall instrument' (fig. 41). Here the semicircle is surmounted by the normal open sights and, while the altitude function has a double shadow square as before, the whole horizontal circle is inscribed within a single square. Digges conceived the instrument as a combination of his 'theodelitus' (or rather half of it) with its double shadow square performing the vertical function, and his geometrical square the horizontal. The result was a truly universal instrument, 'seruing most commodiously for all manner mensurations' (Digges, 1571, sig. I).

Universality was the common goal of the exotic instruments. If the altazimuth was one generic form, others were the recipiangle and the trigonometer, both of which depended on forming, in a small com-

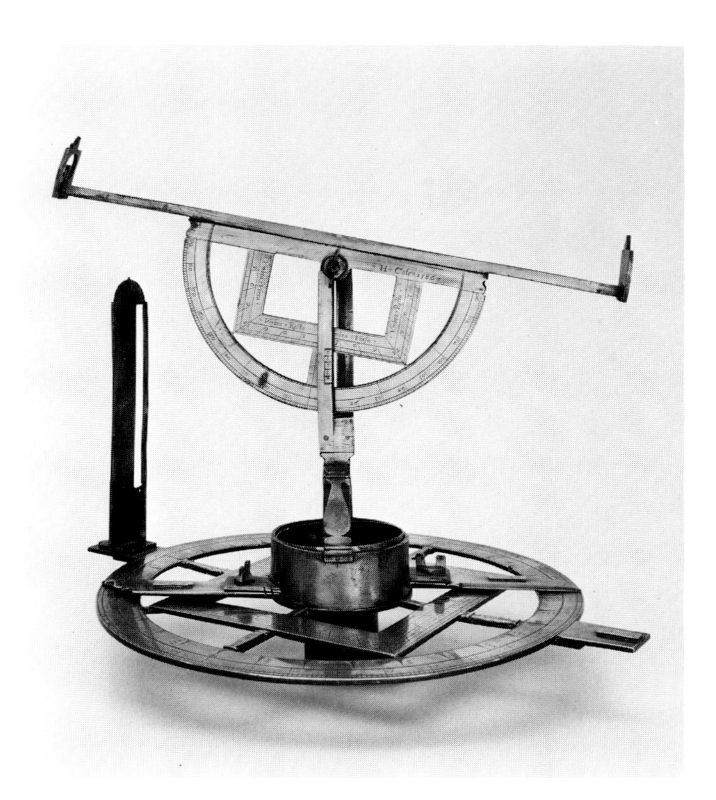

41 (*left*) An altazimuth theodolite by Humphrey Cole, 1586, based on the 'topographical instrument' of Thomas Digges, except that Cole has the shadow square in the base inscribed within the circle rather than circumscribing it. The vertical semicircle, divided to degrees for altitudes, has an inscribed double shadow square. Oxford, Museum of the History of Science.

42 A trigonometer by Danfrie, late 16th century, length 375mm. Sights are taken at either end of a base line, so that a triangle, similar to the one on the ground, is formed by the instrument. Danfrie described the instrument in 1597 and has engraved this example 'Cet Instrument est dit Trigomettre'.

pass, a triangle similar (in the geometrical sense) to the one required on the ground. Textbooks describing many such instruments appear in the late sixteenth and early seventeenth centuries, but are too numerous to be mentioned individually. The instruments include the 'familiar staffe' of John Blagrave, the 'trigonometre' (fig. 42) of Philippe Danfrie (*fl.*1597), the 'geodeticall staffe' of Arthur Hopton (1588–1614), the 'Henry-metre' of Henry de Suberville (*fl.*1598) and the recipiangles of Leonhard Zubler (1563–1609) and of Joost Bürgi. While there are many textbook titles, it is significant that instruments are very rare.

The promoters of such universal gadgets met, it seems, with the

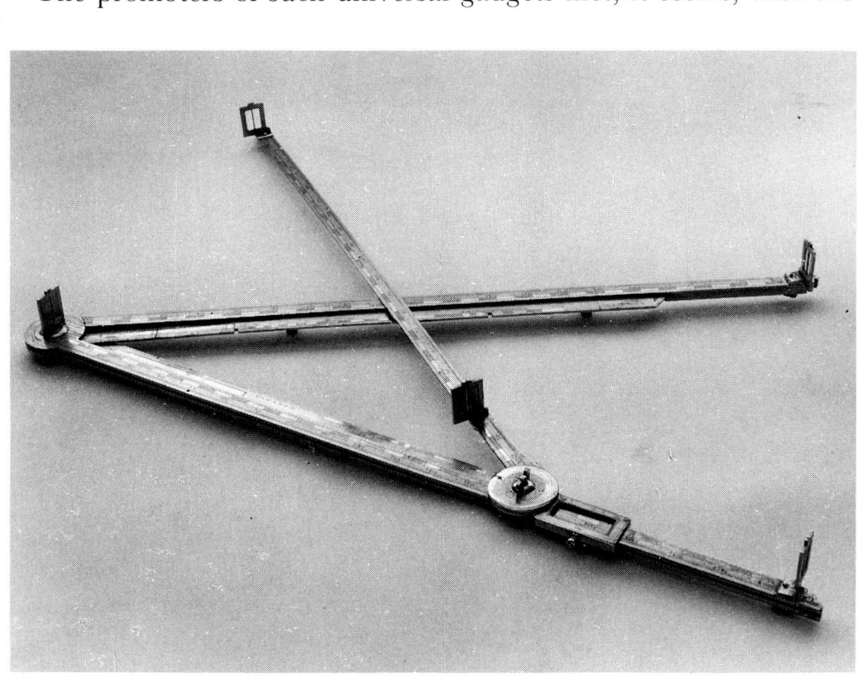

43 Triangulation instrument by Erasmus Habermel, late 16th century. Oxford, Museum of the History of Science.

44 A triangulation instrument in action. It is typical of illustrations of these instruments that they are shown used in battle, in this case to decide the range of the target, which – for good measure – is on the far side of an estuary. The instrument is ascribed to Joost Bürgi, and an example by Bürgi is preserved in the Museum of the History of Science, Oxford. (Bramer, 1648)

implacable resistance of the hard-headed surveyors. Instruments such as the circumferentor and the plane table, better suited to the real needs of the surveyor and with no direct ancestry in astronomy, began to be recommended instead. By 1616, when a measure of sobriety was drawing surveying instrumentation back to Earth from its brief imaginative flight, Aaron Rathborne's more prosaic textbook, *The surveyor*, mentioned such instruments only in passing: 'Nor will I exclude or wholly neglect the Familiar Staffe of M. Iohn Blagraue, and the Geodeticall staffe and Topographicall glasse of M. Arthur Hopton (though now together dead) . . . ' (Rathborne, 1616, p.123).

Practical instruments

Two instruments, which begin to appear in late sixteenth- and early seventeenth-century textbooks, represent the response of practical surveying to the advent of geometry. These were the plane table and the circumferentor, both instruments not handed on to surveying from another discipline, but originating in response to the needs and methods of surveyors.

The plane table (fig. 45), indeed, was resisted by the mathematicians. After being described in 1590 in a textbook by Cyprian Lucar (b.1544), it was grudgingly introduced by Thomas Digges into the 1591 edition of *Pantometria* as 'a Platting Instrument for such as are ignorant of Arithmetical Calculations', and 'an Instrument onely for the ignorante and unlearned, that have no knowledge of Noumbers' (Digges, 1591, p.55).

It consisted simply of a flat board, generally square, used in a horizontal plane, and mounted on a universal joint carried on a tripod stand. A sheet of paper was either pinned to the board or held in place by a close-fitting surround, which might carry graduations to permit the measurement of bearings. Such measurement was not essential, however, as a detached alidade rested freely on the paper, so that bearings could be marked directly onto the paper (one edge

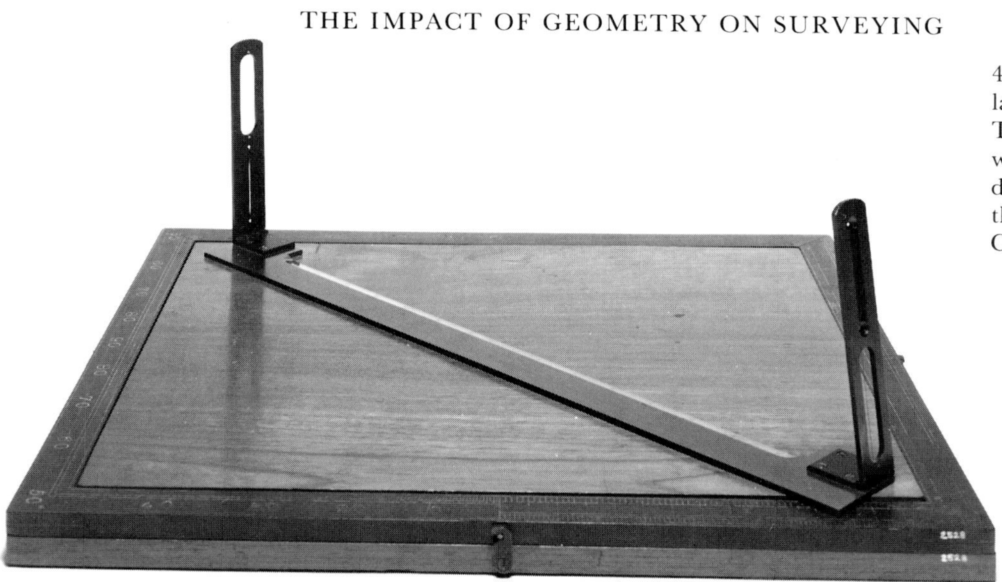

45 Plane table and alidade by Cary, late 19th century, 432 by 379mm. The sights are double slit and window, for sighting in both directions, and they are aligned with the beveled edge of the rule. Cambridge, Whipple Museum.

was usually bevelled to form a rule) and by moving the table from one station to another a map could be drawn up 'in the field'. This obviated the need to measure angles and subsequently 'protract' or 'lay down' these measurements in the office. The popularity of the plane table derived from this simplification.

And its popularity is clear from subsequent textbooks. It was described again by John Norden in his *Surveyor's dialogue* of 1607, and very fully by Arthur Hopton in his *Topographicall glasse* of 1611. Hopton even makes a joke of the table's success: 'it hath begot itselfe a wonderfull affection of the vulgar, whereby they vainly thinke noe worke doth relish well unlesse it bee served upon this plaine Table' (Hopton, 1611, p.102). The danger, of course, from the mathematician's point of view, was that since an understanding of geometry was not essential to using the plane table, the technical basis of this emerging mathematical science might be compromised.

However, the plane table is accepted in Rathborne's *Surveyor* as a standard surveying instrument, and is illustrated on his title-page (fig. 46). He, too, is sensitive to the need to preserve professional standards. The use of the table is apparently so straightforward that 'the multitude of simple and ignorant people (vsing, or rather abusing, that good plaine Instrument, called the Plaine Table)', is in danger of thinking that by merely observing a surveyor at work can its use be mastered (Rathborne, 1616, preface). It must be included, however, since Rathborne's is a realistic, practical work. Gone are the rather fanciful illustrations, typical of earlier texts and especially those promoting recipiangles and the like, of instruments used in the thick of battle. Gone too are the exotic instruments themselves. It is true that the altazimuth theodolite is illustrated on Rathbourne's title-page, but it is not described in detail in the text. Here the reader is referred to Digges's *Pantometria* and to the instrument-maker Elias Allen (*fl.*1606–54), who could provide one if required. It is, wrote Rathborne, 'not altogether so fitting and commodious as the rest before named, by reason of the multiplicity of Diuisions therein contained, which will bee so much the more troublesome in vse and protraction' (Rathborne, 1616, p.124). (Rathborne is one of the very rare examples of writers before the eighteenth century who use 'theodelite' to refer to an altazimuth instrument.)

The other instruments recommended by Rathborne without such qualification, were the simple theodolite (which he calls the 'peractor'), the plane table and the circumferentor. These were supplemented by a chain for linear measure – the 'decimal chain', whose invention he claimed for himself, with each perch divided into 100 parts. The circumferentor (fig. 47) was a form of surveying compass – a magnetic compass with two fixed sights. The essential difference between the simple theodolite and the circumferentor is that bear-

46 Title-page from Aaron Rathborne, *The surveyor* (London, 1616), illustrating the altazimuth theodolite and the plane table.

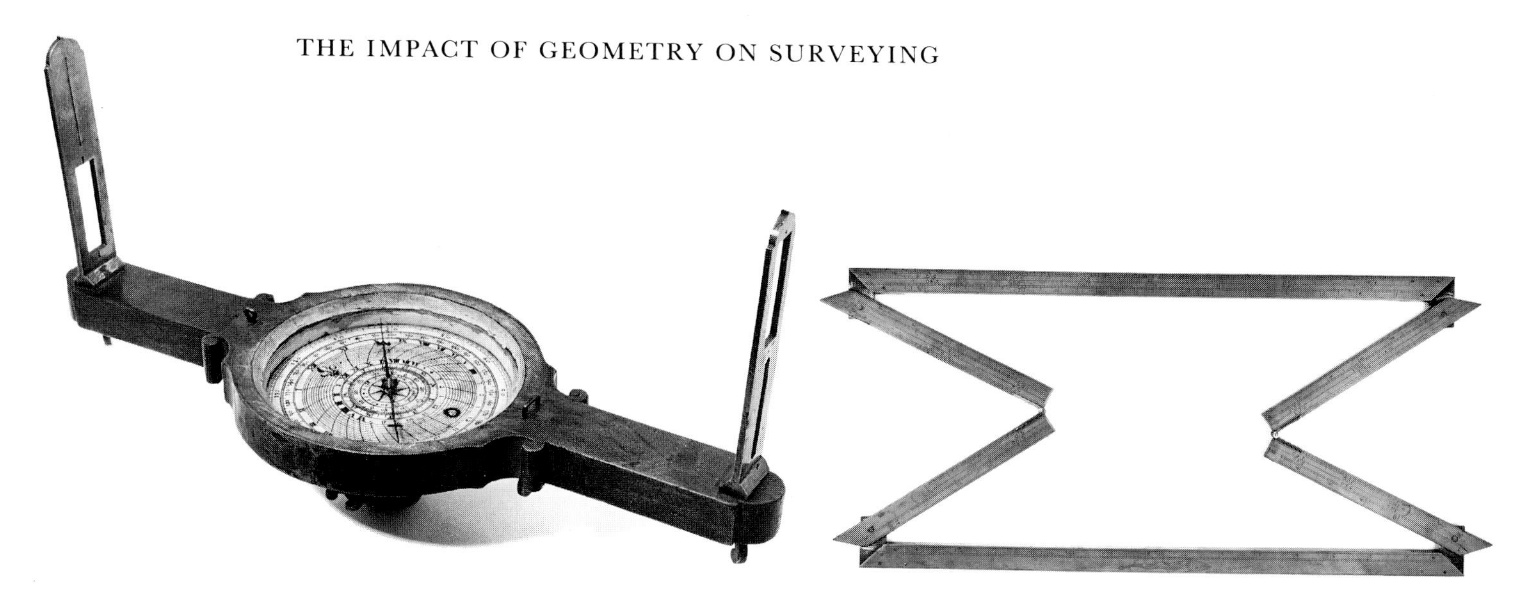

ings are indicated with the former by an alidade on a divided circle, with the latter by a magnetic needle (or, in early forms, by a compass card mounted on the needle). In the former case, the circle is kept fixed while the pointer (alidade) moves; in the latter, the circle moves while the pointer (magnetic needle) is stationary. This distinction was to become important in the development of different types of surveying instruments, suited to different working contexts. While the term 'circumferentor' was consistently used to refer to a surveying compass, a standard form of instrument was not established until the early eighteenth century.

In France, an alternative to the simple theodolite was the 'graphometer' (fig. 49), whose design was published by the instrument-maker Philippe Danfrie in 1597. Essentially half a simple theodolite, the instrument was a divided semicircle with a centrally-pivoted alidade. A small compass was included, for orientating the instrument when taking bearings. Danfrie's instrument could be set in a vertical plane and levelled by a plumb-bob. It could then be used either to take altitudes, using the alidade, or as a level, using fixed sights at either end of the diameter of the semicircle. Later graphometers, popular on the Continent rather than in England, dispensed with the vertical facility, but incorporated a larger compass, with a divided circle, within the semicircle. The fixed sights then allowed it to be used as a circumferentor.

47 (*left*) A wooden Irish circumferentor, with a printed compass card signed 'W. R. Dublin', 1667, length 355mm. This instrument is also a sundial; with the south point of the compass directed to the Sun, the time can be read off the line on the card appropriate to the solar declination, where it meets the needle. Oxford, Museum of the History of Science.

48 (*right*) Folding plane table frame, used to hold the paper in place and marked with degree and linear scales, 18th century, 416 by 341mm. Such frames seem to survive more commonly that the tables themselves. Cambridge, Whipple Museum.

A working compromise

As we have seen, the textbook of Aaron Rathborne (1572–1618) manifests a more realistic attitude adopted towards the needs of the working surveyor. This process advances further in William Leybourn's standard and popular book, *The compleat surveyor*, which first appeared in 1653 and went through five editions between then and 1722. For Leybourn (1626–1716), the 'theodolite' was the simple theodolite and he specifically declined to discuss Digges's 'topographical instrument'. For him, as with Rathborne in practice, the principal English instruments were the simple theodolite, the circum-

49 An example of the original design of graphometer, made by the inventor Danfrie, *c.*1600, radius 140mm. Greenwich, National Maritime Museum.

ferentor and the plane table. He dealt also with Rathborne's decimal chain and with the new chain of Edmund Gunter (1581–1626), where 100 links comprised 66 feet, so that 10 square chains made an acre.

The instrumentation of surveying had changed profoundly during the century between Digges's *Tectonicon* and Leybourn's *Compleat surveyor*, and the change reflected a new view of the surveyor and his domain. Pressure for change had come from the mathematicians, convinced that surveying should take its place among the mathematical sciences. Their vision had been tempered by the realities of surveying practice, but the resulting compromise established a stable group of useful, working instruments. The compromise was admirably explained by Leybourn:

> The particular description of the severall Instruments, that have from time to time been invented for the practice of Surveying, would make a Treatise of it self. . . To omit therefore the description of the Topographical Instrument of Mr. Leonard Diggs, the Familiar Staff of Mr. John Blagrave, the Geodetical Staff and Topographical Glass of Mr. Arthur Hopton, with divers other Instruments invented and published by Gemma Frisius, Orontius, Clavius, Stoflerus, and others; I shall immediately begin with the description of those which are the ground and foundation of all the rest, and are now the onely Instruments in most esteem amongst Surveyors; and those are chiefly these three, the [simple] Theodolite, the Circumferentor, and the Plain Table.
>
> (Leybourn, 1679, p.41)

4

The New Science of the Seventeenth Century

HE SEVENTEENTH CENTURY DESERVES special consideration. It was in this period that mathematical science exercised its most profound influence on natural philosophy: the techniques of such mathematical disciplines as astronomy, navigation and surveying served as models for new ways of understanding and investigating the natural world. We have seen that the mathematical sciences had experienced a notable revival in the early modern period, and that much of this interest had stemmed from practical motives, stimulated by the requirements of navigation and surveying, together with other disciplines, such as engineering and architecture. We have also noted the distinction between this activity and natural philosophy. It was during the seventeenth century that the natural world came to be regarded as a machine – based on mathematical principles, effected, usually at the microscopical level, by mechanical means. A method of investigating the natural world, appropriate to its mechanical nature, was also evolved, namely an experimental method, whose most potent symbol was the scientific instrument, previously the trademark of mathematical science.

The subject is much too large for a full treatment here, but some partial aspects are important for our immediate concern with mathematical instruments, and they will give an impression of the whole. The topics we shall consider – in turn, though we shall find them closely interrelated – are terrestrial magnetism and the longitude problem, the Mercator projection, the advent of optical instruments and the invention and application of telescopic sights, and the changing institutional and social organization of science.

Navigational magnetism

It was noted in Chapter 2 that by the early sixteenth century mariners had observed that magnetic variation – the deviation of the magnetic needle from true north – was not fixed, but varied from place to place. This was of great concern to them, most immediately because of their dependence on the magnetic steering compass. If it was accepted that variation was a real natural phenomenon, and not merely an instrumental error as some argued, it was no longer sufficient to apply a standard correction. Rather, variation had to be discovered for each locality, either by measuring it on the spot, or by deriving it from some geomagnetic theory.

Thus there appeared in the sixteenth century an altogether new kind of magnetic compass for the use of seamen – not a steering compass as such, but a more specialized instrument, the 'variation' or 'azimuth' compass (figs 50, 52, 75, 97, 141). There were a number of forms, each incorporating a device for establishing an alignment with a heavenly body – sights for targeting the stars or the Sun, or a 'style' (usually a string or wire) for casting a shadow. Comparison between this alignment and the magnetic compass built into the instrument yielded a measurement of variation.

True north could be established in theory by a variety of techniques – most straightforwardly by a direct sighting on the Pole Star, or on the Sun or a star at maximum or minimum altitude. This could be compared directly with the magnetic bearing. Alternatively, the Sun's magnetic bearing could be taken at the same altitude in the forenoon and afternoon, with the mean position giving the variation. A similar 'equal altitudes' technique could be used with a star. An 'amplitude compass', on the other hand, required only one observation; it was used to measure solar amplitude – the Sun's azimuth at sunrise or sunset – and the compass measurement could be compared with the true amplitude, given by tables according to latitude and date.

Some steering compasses – for example, many Dutch ones from the seventeenth century onwards – allowed the needle's position to be adjusted with respect to the card, according to the measured variation. It was more usual, however, to apply the correction arithmetically than instrumentally.

With seamen increasingly aware of the risks occasioned by ignorance of magnetic variation, and instrument-makers responding to the pressing need to measure it accurately, it is not surprising that

50 Azimuth compass by J. Fowler, *c.*1720, diameter 305mm. With this form the index is aligned with the Sun using the shadow cast by the string on the target line. Greenwich, National Maritime Museum.

developments in geomagnetic study in the later sixteenth century originated in this practical context. In 1581 Robert Norman (*fl.*1560–96), a compass-maker and, in his own words, 'an unlearned Mechanician', published a book called *The newe attractive*. This was a wide-ranging study of magnetism which included an investigation of magnetic inclination or dip. Inclination first attracted Norman's attention because, to a maker, it was a practical nuisance: customers complained when needles would not rest level. He devised an instrument which would measure this inclination of the needle to the horizontal. This was his dip circle (fig. 51), made from the back of an astrolabe, with a magnetic needle pivoted centrally on a horizontal axis.

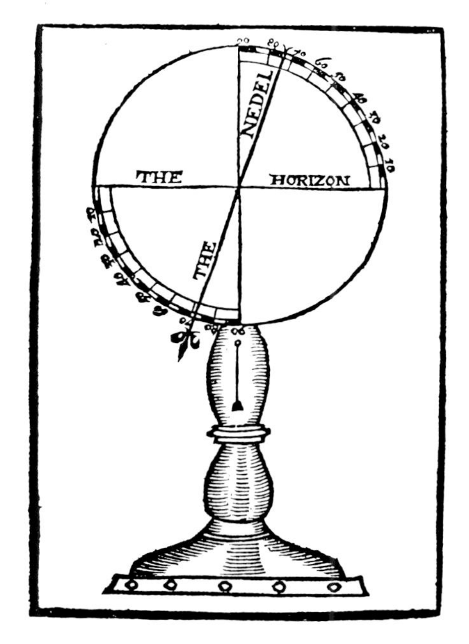

51 Robert Norman's dip circle: a divided circle with a magnetic needle on a horizontal pivot. A plumb-line has been added to the stand. (Norman, 1581)

Published together with Norman's book was a complementary account of variation, William Borough's *Discours of the variation*. Like Norman, Borough illustrated instruments – in this case, several designs for variation compasses (fig. 52) – and he included careful instructions for determining magnetic variation.

An interesting feature of these books is the emphasis their authors placed on experience, experiment and measurement rather than authority. This is especially true of Robert Norman, who wrote that experiments were the basis of all his work. Further, his dip circle is a very early example of an instrument devised for an investigation in natural philosophy; instruments had previously been confined to the mathematical sciences. Instruments of natural philosophy would become, in the seventeenth century, the most striking indication of new experimental and mechanical approaches to the natural world. It is no accident that such techniques first appear where a practical mathematical science – navigation – becomes involved with a question of natural philosophy.

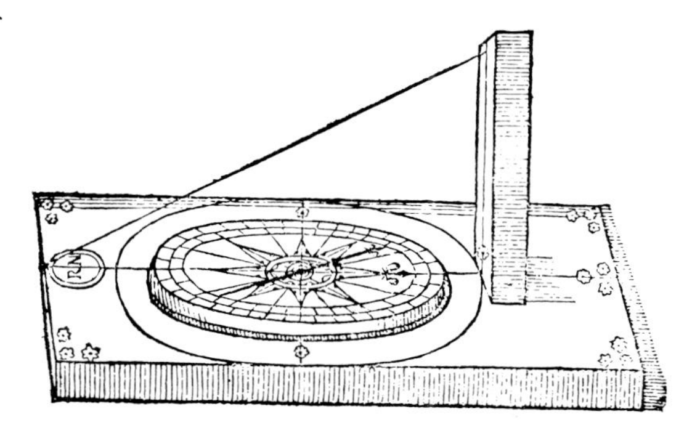
52 A variation compass illustrated by William Borough. When the shadow of the string is made to coincide with the target line, the Sun's magnetic azimuth can be read off the compass in the base. The inscription 'RN' indicates that the instrument has been made by Robert Norman. (Borough, 1581)

This practical and instrumental approach is found in other works on magnetism in the period, written by Thomas Blundeville, William Barlow and Edward Wright, and William Gilbert's *De magnete* of 1600 has long been hailed as an influential early instance of the thoroughgoing use of an experimental method. In this respect, it is properly located within current concerns in navigation.

As common ground between mathematical science and natural philosophy, magnetic variation continued, in the seventeenth century, to provide a meeting-point, where such techniques as practical manipulative investigation and the use of instruments could be transferred into natural philosophy. This role was also played by common interest in a navigational problem that involved geomagnetism but extended beyond it, the problem of longitude.

The longitude problem

A geomagnetic solution to the longitude problem was the one most favoured in England in the later sixteenth century. John Dee had supported the idea that variation measurements could be used to determine longitude, and in 1576 the famous navigator Sir Humphrey Gilbert (?1539–83) announced that he knew of such a method. Such speculations had been prompted by the account of variation given by Cortés in his *Breve compendio*, which was available in English from 1561.

According to Cortés, there was a fixed attractive point, located in the heavens at some distance from the pole, towards which magnetic needles were drawn. This was a theoretical affirmation that variation was not simply an instrumental error, but a real phenomenon of nature. It also predicted a definite relation between latitude, longitude and variation. By finding the two measurable quantities in this relation, the third could be discovered.

By the end of the century, such grand geomagnetic accounts had become unfashionable. Seamen found that in its changes from place to place variation showed scant regard for geometrical functions. Borough had been cautious and Bourne sceptical. In 1599 Wright published an English translation of *The haven-finding art* of the Dutch mathematician and engineer Simon Stevin (1548–1620). Stevin advocated plotting variation all over the globe in an empirical fashion, and using the pattern thus established as the basis of position finding at sea. But this pattern was not expected to bear any particular relation to longitude. This sceptical attitude was reinforced by Gilbert's *De magnete*, with which Wright was closely associated, where variation was regarded merely as local deviation from the magnetic norm, occasioned by the irregularities of the Earth's surface.

By now, the English had established a special interest in magnetism, appropriate to their seafaring ambitions. They had also established an institution where mathematical science would be a particular concern, and it was here – in Gresham College in London – that the magnetic enterprise was pursued further.

The College had been founded in 1597, through an endowment of the celebrated financier Sir Thomas Gresham (*c.*1519–79). Here were established the first mathematical professorships in England – there were none in the universities – and each professor's principal duty was to deliver weekly public lectures. Contrary to university practice, they were to be given in English as well as in Latin.

The statutes governing the mathematical lectures – geometry and astronomy – were appropriate to lectures delivered in English to a public audience in London. They were intended to be relevant to the needs and interests of the practical mathematical community, and the Geometry Professor was to teach arithmetic and practical, as well as theoretical, geometry. The Astronomy Professor was to include navigation and the use of instruments.

While the statutes were not always strictly applied, the spirit of the College and the interests of the early mathematical professors were in keeping with these aspirations. The first Professor of Geometry was Henry Briggs (1561–1630), who introduced into his lectures that important aid to practical arithmetic – logarithms, recently invented by John Napier (1550–1617). Among his associates were Edward Wright and William Gilbert (1540–1603), and also Thomas Blundeville and William Oughtred (1575–1660). Another friend was Edmund Gunter, who became Professor of Astronomy at Gresham in 1619. Gunter had extensive interests in navigation and other aspects of practical mathematics, and his name is associated with a number of practical instruments: Gunter's chain (which became the common form of surveyor's chain), Gunter's sector (a calculating instrument particularly adapted to navigation), Gunter's rule (another navigational calculator, incorporating logarithmic functions), Gunter's staff (a cross-staff that served also for calculation) and Gunter's quadrant (an horary quadrant with features for use at sea).

Gunter measured the magnetic variation in 1622, purposefully at a location used by William Borough, and noted a significant discrepancy between his measurement and that announced by Borough in 1581. A further determination in 1634 by Gunter's successor as Gresham Professor of Astronomy, Henry Gellibrand (1597–1636), indicated a continuing change in magnetic variation over time, which he announced in 1635 as 'that abstruse and admirable variation of the variation' (Gellibrand, 1635).

While secular changes in magnetic variation made any application to longitude-finding almost impossibly complex, it did not extinguish interest in such schemes. Strangely perhaps, interest was rekindled. There was, after all, little excitement in the prospect of measuring, all over the globe, a variation uniquely determined by local factors in every case. We have seen magnetic variation as a meeting-point between practical mathematical science and natural philosophy. Now natural philosophical interest was heightened, for secular changes might serve as clues to more obscure underlying causes than irregularities in the Earth.

An extensive and detailed survey by Edmond Halley (*c.*1656–1742) of magnetic variation in the Atlantic at the end of the century, was aimed at just such a longitude method, based on a sophisticated geomagnetic theory, which incorporated time as one variable. Robert Hooke (1635–1702) also had devised a theory in the 1670s. More immediately, however, Gellibrand's announcement prompted the mathematical practitioner Henry Bond (*c.*1600–78) to evolve a theory of variation changes and to announce predictions in the annual *Seaman's kalender*, which he was then editing. The success of his predictions, together with his longitude claims, aroused interest and played, as we shall see, an important part in the prehistory of the Royal Observatory, founded in 1675 to provide quite a different solution to the longitude problem. This rival – the lunar method – had thus received an official endorsement even before Bond published, with premature optimism, *The longitude found* in 1676.

While the lunar method was only one of the rivals to longitude by geomagnetism, common to each of the proposed alternatives was some method of determining standard time. Differences in local time are equivalent to differences in longitude: for every 15° movement in longitude east or west, local time by the Sun is different by one hour. Thus if a seaman can discover the time at some meridian conventionally adopted as zero, comparison with his own local time constitutes a longitude measurement.

The lunar method (figs 53, 54) proposed to utilize the relatively fast motion of the Moon through the fixed stars as a great celestial clock, visible throughout the world. The most commonly cited method of applying the lunar motion was by 'lunar distances', proposed by Werner in his 1514 Nuremberg edition of Ptolemy's *Geography*, and repeated in Apianus' *Cosmographia*. Astronomers would provide predictions of the Moon's position with respect to designated stars, and supply them to navigators in the form of tables calculated for time at some standard meridian. Navigators would consult the celestial clock by measuring the Moon's position, making a number of necessary corrections, and referring to the tables.

While the idea, like other longitude solutions of the time, was sound in theory, the means of applying it were almost entirely lacking. The complex lunar motion was inadequately understood, the positions of the fixed stars had not been recorded with sufficient precision, and there was no instrument for making the observation at sea with the required accuracy. A further obstacle was the general mathematical competence of seamen. The process of reducing the raw observations to a useful lunar distance was laborious and demanding. For example, corrections had to be made for parallax (the Moon's apparent position with respect to the stars is slightly different according to the observer's position on Earth, assumed to be known approximately) and for atmospheric refraction (which varied

53 Title-page from J. Werner and P. Apianus, *Introductio geographica Petri Apiani in Doctissimas Verneri annotationes* (Ingolstadt, 1533), showing the instruction of a mariner in the use of the cross-staff to find longitude by lunar distance, and the instrument's application to surveying.

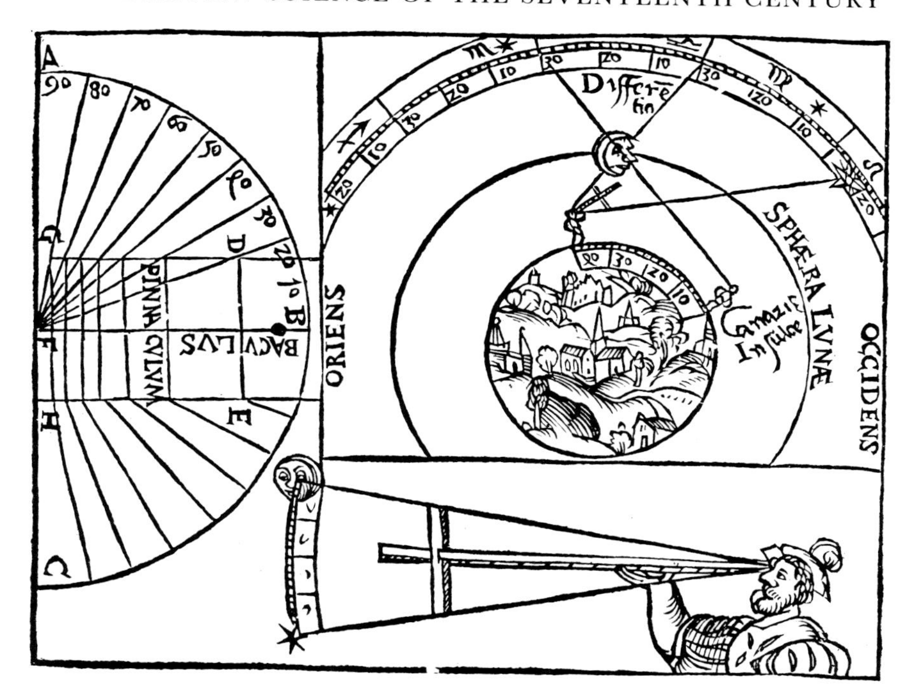

54 The cross-staff used (upper right-hand diagram) to measure lunar distance for longitude determination. The division of the staff is illustrated in the left-hand diagram, and the method of use in the lower right. (Apianus, 1539)

with the altitude of the observation, according to the observer's latitude). It would be some time, however, before the method's progress was sufficient to encounter the limited mathematical literacy of seamen.

Isolated attempts to apply a lunar method are recorded, but were generally land-based and employed the special and simplified case of simultaneous observations of a lunar eclipse, where parallax and refraction are not involved. Gellibrand, for example, took part in one such experiment, an account of which was published in 1633 as an appendix to *The strange and dangerous voyage of Captaine Thomas James*. James's primary purpose was to search for a north-west passage to the Indies, but there were other mathematical and philosophical goals, for the instruments he carried were appropriate to what we might regard as a scientific expedition: they included six variation compasses 'besides some doozens of others, more common' (James, 1633, sig. Qv), cross-staves (which included large ones of 6 ft and 7 ft in length), quadrants (including one of 4 ft radius, very large for a portable instrument) and many other instruments for navigation and surveying.

James and Gellibrand had arranged the simultaneous observation of a lunar eclipse – James at Hudson's Bay and Gellibrand at Gresham College. Gellibrand had a considerable instrument – a quadrant of 6 ft radius, divided to minutes of arc – which is a useful indication of resources at Gresham in 1631. He tells us that James was 'in his discouery for the N.W. at the bottom of the Bay, being his wintering place, and called by the name of Charloton [sic: he generally gives 'Charleton']' (Gellibrand, 1633, sig. Rv). In fact, James tried two methods of determining standard time: by lunar eclipse and by the Moon's passage across his meridian. In the latter, more difficult, case Gellibrand subsequently calculated a longitude of

78° 30′. From the eclipse observation, he concluded 79° 30′, which is the modern value for Charlton Island at the southern extremity of Hudson's Bay.

Eclipses are, of course, too infrequent to be much use at sea, quite apart from the difficulties of observing on board ship, and the lunar distance method remained the goal of navigational theorists. The method was well known among mathematicians, through the popular *Cosmographia* of Apianus and Gemma Frisius. It is treated in the earliest English manual of navigation, Bourne's *Regiment for the sea* of 1574, and also in the navigation section of his *Almanacke and prognostication for three years* of 1571. Bourne concluded, however, that it was not practical and he advised seamen to stick to their traditional method of dead-reckoning:

> I woulde not any Sea men shoulde be of that opinion that they mighte get anye Longitude with instrumentes . . . but (according to their accustomed manner) let them keepe a perfite accompt and reckening of the way of their shippe, whether the shippe goeth to lewards or makith hir way good, considering alwayes what thinges by against them or with them: as tides, currents, winds, or such like. (Bourne, 1963, pp. 239–40)

Although it had long been familiar to specialists, an occasion which brought the lunar method to wider public notice had important consequences and still affects the daily lives of almost everyone in the world.

In 1674 a Frenchman at court was promoting the claims of a lunar method to Charles II. Considerable financial rewards were expected to flow from a longitude solution. Philip II and Philip III of Spain had earlier offered prizes, and their examples were followed by other European powers. Charles had already appointed a Royal Commission to investigate the claims of Henry Bond, and many of the same group were asked to consider the new proposal. They included Christopher Wren (1632–1723) and Robert Hooke, and the mathematician and Surveyor-General of the Ordnance, Sir Jonas Moore (1627–79), was also closely involved. It was through Moore that John Flamsteed, an able young astronomer and Moore's protégé, attended meetings of the Commission and came to impress on Charles the well-known deficiencies in the lunar method and their source in the general inadequacies of precision astronomy. Charles was moved to supply these inadequacies by building an observatory and appointing Flamsteed as his astronomer. According to Flamsteed's own account of the siting of the Observatory, 'Sir Christopher Wren mentioning Greenwich Hill, it was resolved on' (Baily, 1835, p.39).

It is important to stress that the Royal Observatory at Greenwich was founded for a very specific purpose, to provide the necessary empirical basis for a lunar solution to the longitude problem. The warrant that authorized payment of John Flamsteed's salary stipulated his duties as:

> forthwith to apply himself with the most exact care and diligence to the rectifying the tables of the motions of the heavens, and the places of the fixed stars, so as to find out the so much-desired longitude of places for perfecting the art of navigation. (Forbes, 1975, vol.1, p.19)

This set a pattern for the work done at Greenwich, with fundamental precision astronomy top of the agenda, and this in turn eventually made Greenwich an appropriate international reference point for longitude and time.

The Observatory was established at a propitious time, for it was able to take advantage of a recent advance in instrumentation which had been discussed in the literature and tried out on individual instruments, but was only beginning to be applied to equipping observatories. This was the replacement of open sights on fixed instruments by telescopic sights, a development of such importance to all forms of precision instruments that it will be considered in a separate section.

The progress of the lunar method will be taken up again in Chapter 8, but we should note here that, while a beginning was made in the seventeenth century towards supplying the necessary stellar and lunar observations, an important step was also taken towards furnishing the navigator with an instrument for use at sea. Robert Hooke and Isaac Newton (1642–1727), apparently independently, devised reflecting instruments whose principles would govern angle measurement at sea from the mid-eighteenth century. The fundamental technique was to bring the two bodies whose angular distance was required, into apparent coincidence by viewing one directly, and the other by reflection in a mirror. The angle between the direct sight (which might be established by the orientation of the instrument) and the mirror (which would then be mounted on an index arm) was half the required angle.

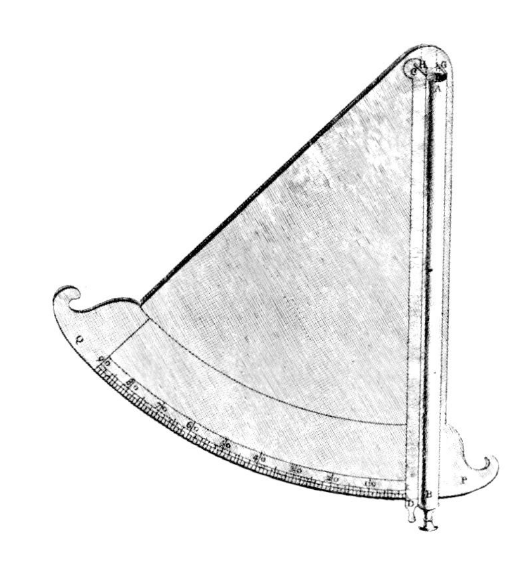

55 Newton's design of reflecting quadrant or octant. The index arm CD rotates the mirror H, so that one object is viewed directly, past the mirror G, and the other after reflection in mirrors H and G. (Newton, 1742)

As early as 1666, Hooke described to the Royal Society a 'perspective . . . for observing the positions and distances of fixed stars from the Moon by reflexion' (Birch 1756, ii, pp. 111–14); Newton showed them a similar instrument (fig. 55) in 1699, later described in print as 'an Instrument for observing the Moon's distance from the Fixed Stars at Sea' (*Philosophical transactions*, 42, p.155). Since the angle moved by the mirror is half that of the reflected ray, a 45–degree arc is shown calibrated to 90°.

The other astronomical method of finding longitude, current in the seventeenth century, was by determining time from the eclipses and occultations of Jupiter's satellites. Four satellites were discovered in 1610 by Galileo (1564–1642), who suggested their application to the longitude. The fairly frequent occasions when apparent contacts occur between satellites, or between a satellite and the body of the planet, might provide a reference clock in the sky, once astronomers had established the necessary tables. This method was pursued in the seventeenth century, first at Gresham College and then at the Paris Observatory, and was actually used on land. The motion of any ship, however, prohibited its use at sea, for at the time a telescope of considerable length was required to establish the instants of contact.

In theory, the most straightforward solution to the longitude problem was to carry a watch set to standard time, which could be compared with the local time. The idea, however, presented the greatest practical difficulties of all. The earliest watches date from around the second quarter of the sixteenth century, and the first written suggestion of the so-called 'chronometer method' for finding longitude dates from the same period. This was made by Gemma Frisius in *De principiis astronomiae cosmographicae* of 1530, and was repeated, for example, in William Cuningham's *Cosmographical glasse* of 1559, which, as we have seen, depended on Gemma also for the principle of triangulation.

To find longitude within one degree – a modest enough ambition – the watch must not err (or rather must not have an unpredicted error) of more than four minutes, and this might be the cumulative error on a voyage lasting months. Sixteenth-century watches, however, had no minute hands, since their expected accuracy was to within a quarter of an hour per day. The gap between the theoretical proposal and the practical reality was immense.

The goal of a longitude watch was the most important stimulus to developments in horology over the next three hundred years. In the seventeenth century, the work of Robert Hooke and of the Dutch mathematician and astronomer Christiaan Huygens (1629–95) led to the invention and improvement of the pendulum clock and of the spring-regulated watch. The pendulum clock may not seem a very practical sea-going instrument, but Huygens did his best to make it seaworthy. He incorporated such features as gimbal mounting to keep the clock level, and 'cheeks' to constrain the pendulum's motion into a cycloidal path, which he showed was performed in equal times regardless of amplitude variations occasioned by the movements of the ship. Seaborne trials were organized in the 1660s, with modest success. Hooke held out more hope for his spring-regulated watches, with whose invention Huygens was independently involved.

The development of horology had important implications for precision astronomy. We have seen that Tycho had clocks for timing his observations; several are illustrated in use with the great mural quadrant. The significant improvement in accuracy and reliability represented by the pendulum clock, brought mechanical clocks into widespread use as astronomical instruments. The apparent daily rotation of the heavens was already used to express positions in terms of 'hour-angles'. Right ascension was measured in hours, minutes and seconds, and the pendulum clock became an instrument for measuring angles expressed in time. A targeting device was, of course, required for determining the instants when the angle 'began' and 'ended'. For this a new instrument emerged in the later seventeenth century, the 'transit instrument', which will be discussed later, as it incorporated a separate development – the invention of the telescopic sight.

Mercator sailing

Sea-charts date from at least the thirteenth century, and were first drawn for the Mediterranean. They reflected the Mediterranean practice of sailing by compass bearing and distance, rather than the later technique of latitude sailing. The chart was covered with a network of bearing or rhumb lines, radiating out from decorated compass or wind roses. The navigator would lay down with a straight rule the direction to be taken, and by tracing the nearest parallel rhumb line back to its rose, discover the appropriate compass bearing.

Within the confines of the Mediterranean, and the limits on accuracy imposed by contemporary navigation, it mattered little that this was a 'plane' chart. In other words, it treated the Earth as a flat surface, though cosmologists had for long known it to be a sphere. Meridians were drawn parallel, though in reality they converge towards the poles. Neither did it matter, in the short term, that no account was taken of magnetic variation, since an appropriate variation was 'built in' when the chart was made.

Atlantic exploration revealed the shortcomings of the plane chart. It was impossible, of course, to match bearings and distances consistently, and courses plotted on the chart were not 'true' – they did not represent the rhumb to be followed for the measured distance.

In 1569 Gerard Mercator published a chart based on a proper projection of the sphere, which was well suited to the needs and customs of seamen. The meridians were parallel, while at the same time a line drawn between two points represented the true bearing to be steered by the helmsman. To achieve this, the latitude scale was varied in proportion to the longitude distortion introduced by paralleled meridians.

The true convergence of the meridians is proportional to the cosine of the latitude. Tables were already available which expressed the lengths of degrees of longitude at different latitudes as proportions of the length of a degree at the equator, and these were used for converting distance sailed east or west into change of longitude. If, however, this cosine function is ignored on the chart and the meridians are drawn parallel, the error introduced into the east-west distances is proportional to the reciprocal to the cosine of the latitude, that is to the secant of the latitude. Mercator increased the spacing between latitude lines the further they are from the equator, introducing a proportional change in the latitude scale that exactly corresponded with the error in longitude. At any point, the latitude and longitude scales correspond, but this common scale, used in calculating distances, varies as the ship travels north or south, and the areas represented on a Mercator chart are correspondingly distorted, the distortion being most marked in high latitudes.

It followed from this variation in scale that all distance calculations required some ability in handling trigonometrical functions. While the distance scale applicable to any course involving change in latitude varied continuously, it was acceptable in practice to apply the scale appropriate to the middle latitude of a straight course on the chart, a practice known as 'mid-latitude sailing'.

Mercator did not explain the mathematical basis of his projection, and this was done by Edward Wright in his *Certain errors in navigation* of 1599. This included a table of 'meridional parts' – the scale variation or spacing of latitude lines along a meridian, which was derived from the table of secants. This was a first step towards substituting

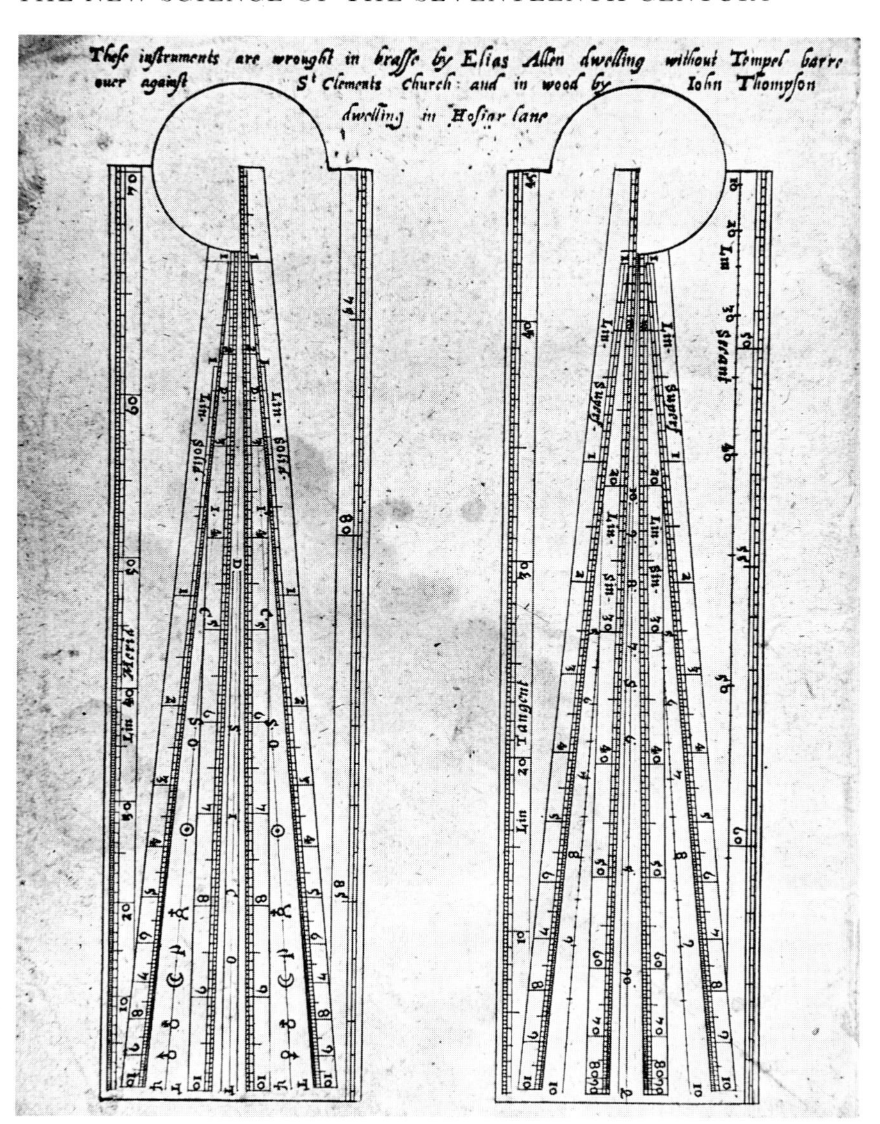

56 Both sides of Gunter's sector, illustrated in a plate almost certainly prepared by Allen himself for Gunter. (Gunter, 1623)

'Mercator sailing' for 'plane sailing', but the mathematics involved would have to be considerably simplified before the technique was accessible to the ordinary navigator. The seaman needed a straightforward, instrumental means of calculation, suitable for use at sea, together with a relatively simple set of rules for applying the instrument to specific problems. Both were provided by Edmund Gunter.

The 'sector' was a folding rule in brass, wood or ivory, engraved with various lines or scales radiating out from the flat joint. It had recently been invented as a calculating instrument of use to surveyors, military engineers and others concerned with practical mathematics. Based on the properties of similar triangles, and used in conjunction with a pair of dividers, it could be used to solve a variety of problems quickly and without calculation.

Gunter adapted the sector to navigational use, and in the *De sectore et radio* (with English text) of 1623, he described its applications. Gunter's sector (fig. 56), in addition to certain features common to other sectors, included lines divided according to trigonometrical

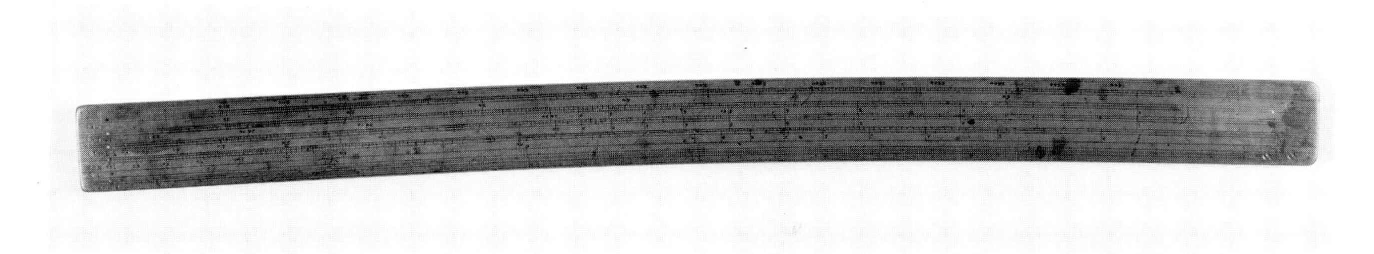

57 Gunter scale or rule, signed 'Robt Yeff fecit', 1712, length 606mm. There are wooden nocturnals signed by Yeff 'in Bristol' at the Science Museum, London, and the National Maritime Museum, Greenwich. Thus Yeff derived at least part of his living from making wooden navigation instruments in a provincial port. Cambridge, Whipple Museum.

functions and a linear scale of meridional parts. His rules explained its use in finding, for example, distances sailed and longitude differences and, in general, in the solution of spherical triangles. With his sector, the solution of navigational problems involving spherical trigonometry was at least feasible, but a second mathematical innovation would simplify such problems still further.

Navigation was one of the first subjects to benefit from the application of logarithms, and here again it was Gunter who supplied the needs of seamen. What they required were logarithmic tables of trigonometrical functions, and these Gunter published in 1620. He went on to incorporate them into a simple instrument ideally suited to the seaman. This was Gunter's scale or rule (fig. 57), at first designed to be engraved on a cross-staff, but basically a rule with logarithmetic functions set out in linear scales. It included lines of logarithms of numbers, as well as of trigonometrical functions, and to it he added a line of meridional parts. Additions and subtractions (equivalent to multiplications and divisions) could be performed mechanically with a pair of compasses (dividers). It was a small step to the logarithmic slide rule, the standard calculation instrument until very recent times.

Even though plane sailing and the plane chart survived in some quarters well into the eighteenth century, the introduction of Mercator sailing was an instructive example of what could be achieved through serious application of mathematics to navigation. A set of problems had been solved, and the techniques of their solution had been translated into truly practical instruments. It illustrates also the increasing importance of the English practical mathematical school.

The telescope and the micrometer

The English were also prominent in the introduction of telescopic sights, probably the single most important technical advance in precision instrumentation in the seventeenth century. Galileo had first brought notoriety to the cosmological use of the telescope, and Johannes Kepler (1571–1630) had devised the particular form that could be adapted for making measurements, but the earliest recorded use of a telescopic sight falls to a considerably more obscure astronomer, the Englishman William Gascoigne (*c.*1612–44).

The optical components of the so-called 'astronomical telescope', whose design was published by Kepler in 1611, comprised, in its most primitive form, a convex objective lens and a convex eyepiece.

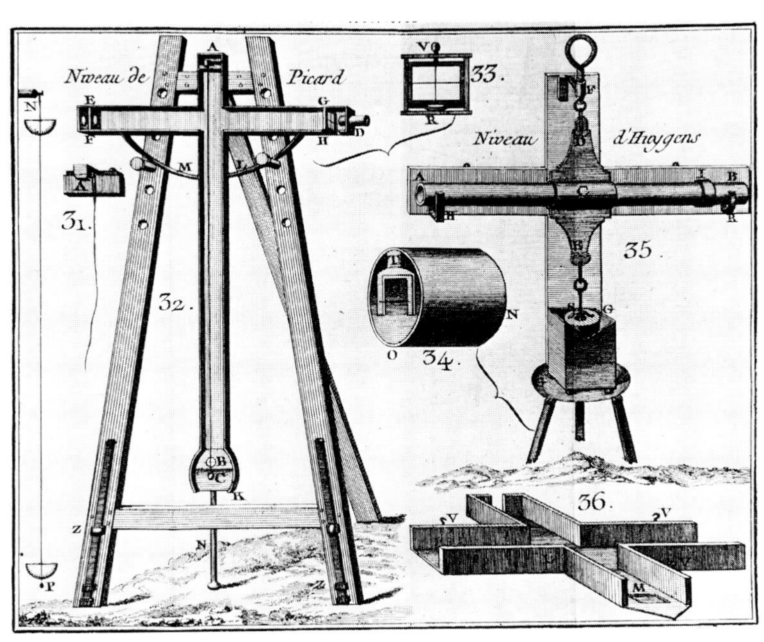

58 (*left*) Moving-wire eyepiece micrometer by John Rowley (?), *c*.1710, length 210mm. The handle is linked by cogs to two opposing screws, which give symmetrical motions to two vertical wires while the pointer registers their displacement. This instrument has always been attributed to Rowley, but the signature is in fact 'John Ronley Fecit'. Cambridge, Whipple Museum.

59 (*right*) Picard's level – a type known as a 'pendulum level', which relies on a heavy plumb-bob to find the vertical line – is on the left. The cross-wires of the telescopic sight are clearly shown. (Picard, 1780)

The function of the objective was to collect an appreciable amount of light from the target object, and to bring it to a focus – to create a real image – within the tube. The eyepiece then magnified this image. Since an image was truly formed in the focal plane of the objective lens, anything introduced into this plane would also be magnified by the eyepiece and would be seen superimposed on the parts of the image left unobscured.

It was on this principle that Gascoigne stumbled in about 1640, after a spider happened to spin a web in the appropriate position. There were two general applications. A pin-point or cross-hair could be used to locate the centre of the field of view, so that the telescope could be accurately aligned on the target object and used as a sight on a traditional astronomical instrument. Alternatively, and for different purposes, a measuring device (fig. 58) could itself be introduced into the focal plane and used to measure small angular distances and apparent diameters.

Gascoigne himself made both applications, and his eyepiece micrometer in particular was a sophisticated instrument, consisting of parallel knife-edges moved by the opposed elements of a double screw, whose displacement was indicated by a pointer moving over a graduated disc.

Gascoigne's work was known to his circle of associates, who included Oughtred, and his techniques were applied in Oxford in the 1650s, notably by Christopher Wren. Wren used micrometers for a variety of astronomical projects, including a measured survey of the Moon, and it was he and later Hooke who pursued Gascoigne's invention in England. Wren also used telescopic sights, for in 1663 he produced at the Royal Society an astronomical instrument equipped with two 6 ft sights. When it was built is not clear, but it was demonstrated in connection with a current Royal Society programme to establish the basis of the lunar distance method for longitude.

In the meantime, Huygens' independent discovery of a form of eyepiece micrometer had been announced in his *Systema Saturnium* of 1659, where he presented his solution to the greatest telescopic puzzle of the age: Saturn was surrounded by a detached, oscillating ring. Huygens introduced a thin, tapered metal plate into the focal plane of the object glass, and its varying width was used as a gauge of apparent angular distance.

Other instances of the introduction of forms of micrometer occur in the 1660s, one of the most influential being the moving-wire micrometer of the French astronomer Adrien Auzout (1622–91), announced in 1667 to both the Royal Society in London and the Académie des Sciences in Paris. In its early form it consisted of one fixed wire and a second wire moved by a screw. Auzout's fellow astronomer Pierre Petit (*c.*1596–1677) also designed related forms of eyepiece micrometer, and Jean Picard (1620–82) contributed to the development of the micrometer and also to the application of telescopic sights to large astronomical instruments and, significantly, to geodetical instruments, including a surveying level (fig. 59).

Thus in the 1660s, two groups were forwarding the use of the micrometer: Wren and Hooke in London and Auzout, Petit and Picard in Paris. It is significant that both groups had access to the organizational structures of newly formed scientific societies, and it would be important that in both cases outlets were found for their achievements in newly established national observatories. Their work marked an important extension to the role of the telescope, which Wren had identified in a letter written in about 1656. He neatly confirms the prevailing distinction between a telescope and an astronomical instrument (on which the reference of this book is based):

> . . . we make the Tube an Astronomical Instrument, to observe to Seconds, by which we take the motions of ♃ Satellites and ♄ moon; and not only draw Pictures of the Moon, as Hevelius has done, but Survey her & give exact maps of her, & discover exactly her various Inclinations, and herein Hevelius's Errors. (Bennett, 1973, p.147)

By the addition of the micrometer, the telescope ('the tube') had been transformed into an astronomical instrument.

Controversy over telescopic sights

Wren's reference to Hevelius' work as the standard by which he judged his own is significant. Johannes Hevelius was the leading observational astronomer during the 1650s and 1660s, a position established by the work to which Wren refers – the *Selenographia* of 1647 with its splendid engravings of the lunar surface. For those trying to introduce and promote the use of telescopic sights, it became important that the most respected observer had set his face resolutely against them.

Hevelius began to construct his observatory (fig. 60) in Danzig (now Gdansk) in the 1640s, and by making his own instruments, he gradually extended the range of his equipment. The observatory was destroyed by fire in 1679 and was never fully reinstated. Hevelius' instruments were carefully illustrated and described in the first part of his *Machina coelestis*, published in 1673. A large number of quadrants and sextants (fig. 61) were provided for astronomical measurements, some being on portable stands and others on fixed mounts, and the principal instruments were between 6 and 9 feet in radius.

Subdivision of primary scales was by diagonals or by the device announced by Pierre Vernier (1584–1638) in 1631 and which bears his name. A secondary scale moving with the near-sight was made equal in length to, say, nine of the primary divisions but divided into ten equal parts. Subdivision of the primary scale into tenths is simply a matter of noting which of the vernier divisions most closely coincides with a primary division. In the case, for example, of an index intersecting a scale at four tenths of the distance between 32 and 33, each division of the vernier scale equals 0.9 of a primary division, and the position of the zero of the vernier is 32.4. Thus the position of the first vernier division will be 32.4 + 0.9 = 33.3; of the second, 32.4 + 1.8 = 34.2; of the third, 32.4 + 2.7 = 35.1; and of the fourth, 32.4 + 3.6 = 36.0. The actual form of the vernier may have changed over the years, but the principle has remained the same.

Seventeen separate astronomical instruments are described in *Machina coelestis*, and they represent a very extensive range. Some were on altazimuth stands and others on ball-and-socket mounts; Hevelius was surprisingly unmoved by the advantages of the equatorial mount. In addition, many telescopes were described, ranging in length from 30 to an extravagant 150 feet. (Long focal length objectives were used in the period to minimize the effects of spherical and chromatic aberration.) Hevelius' concern with the development of the telescope and even with the use of eyepiece micrometers seems strangely at odds with his firm opposition to telescopic sights on traditional instruments.

The sights employed by Hevelius were of the type developed by Tycho Brahe, with cylindrical fore-sights and double-slit near-

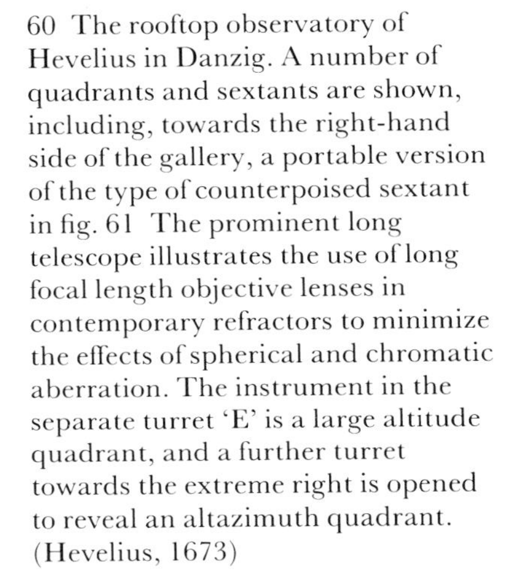

60 The rooftop observatory of Hevelius in Danzig. A number of quadrants and sextants are shown, including, towards the right-hand side of the gallery, a portable version of the type of counterpoised sextant in fig. 61 The prominent long telescope illustrates the use of long focal length objective lenses in contemporary refractors to minimize the effects of spherical and chromatic aberration. The instrument in the separate turret 'E' is a large altitude quadrant, and a further turret towards the extreme right is opened to reveal an altazimuth quadrant. (Hevelius, 1673)

61 Hevelius observing with a large brass sextant, assisted by his wife Elizabetha. It is mounted on a pillar, with a ball-and-socket joint, and is attached to ropes leading over pullies to counterpoising weights. Note the sights of the type developed by Tycho Brahe – a cylindrical fore-sight and double-slit near-sight. (Hevelius, 1673)

sights, sometimes mounted on an alidade, sometimes moving independently on the limb. Although they were remarkably effective in Hevelius' hands, the overall character of the instruments was old-fashioned, and it is understandable that those who sought to establish a new approach to astronomical instrumentation felt obliged to indicate his neglect of telescopic sights.

Hooke was Hevelius' principal adversary in the subsequent controversy. They had already been in correspondence, with Hooke trying to impress upon Hevelius the advantages of telescopic sights, so that the latter had included a defence of his open sights in the *Machina coelestis*. Hooke replied with vigour in his *Animadversions on the first part of the Machina Coelestis of . . . Johannes Hevelius*, published only the following year in 1674. It was probably his most important publication on instruments. In the same year, he had published *An attempt to prove the motion of the Earth from observations*, where he described a 36 ft zenith telescope (i.e. one mounted vertically) with an eyepiece micrometer, set up in his rooms at Gresham College, in an effort to detect the apparent annual motion of the stars predicted by the Copernican theory. In *Animadversions*, he looked forward in time to the kind of instruments he wished to see used to re-establish the observational basis of precision astronomy.

Hooke's fundamental argument against Hevelius was that there was a limit to the resolving power of the human eye – the ability to distinguish an angular separation of less than a minute was very rare, and of half a minute impossible.

> And this being proved, what will become of all the machinations and contrivances for greater instruments, to shew the Divisions of single or double Seconds? May not single minutes, nay half minutes, by the help of Diagonal Divisions, be sufficiently distinguished in an instrument of three foot Radius? What need is there then of all the other cumber? . . . if the eye cannot distinguish a smaller object then appears within the angle of half a minute, 'tis not possible to make any observation more accurate, be the instrument never so large. (Hooke, 1674, pp. 7–8.)

Hooke's answer, of course, was to assist the eye with telescopic sights. An artificial extension to the natural resolving power would then permit new levels of accuracy. He too was surprised that, of all people, Hevelius should be opposed to their use: 'I cannot choose but wonder why he should be of that opinion, who hath not been less exercised in the use of the Telescope, then any present in Europe' (*ibid*, p.43). Part of the answer may lie in the distinction, current in Hevelius' early career, between telescopes and astronomical instruments.

Hooke went on to describe a single instrument (fig. 62) that incorporated many of his ideas. It was a very splendid conception – an equatorial quadrant, with two telescopic sights arranged to superimpose their images and to allow a single observer to take both sights at

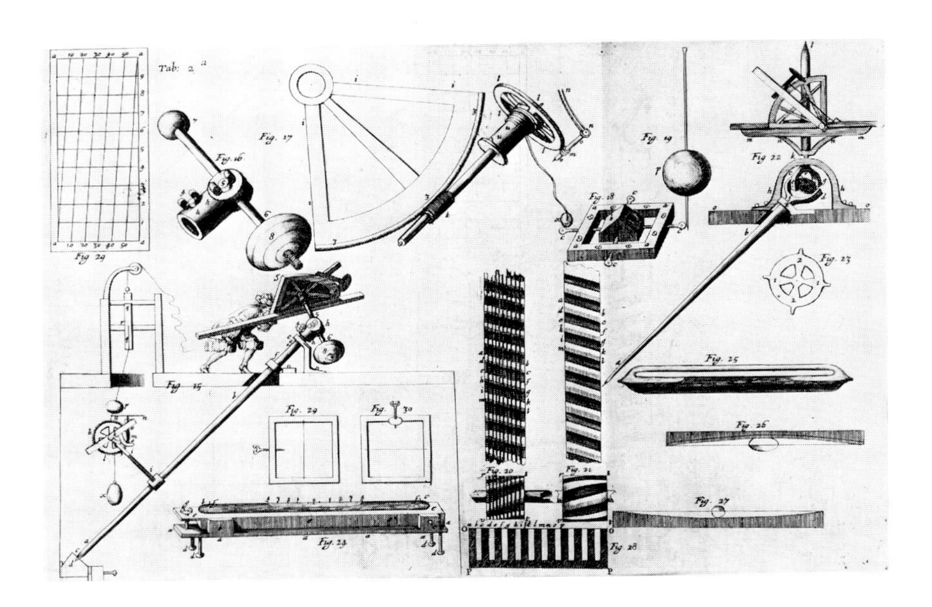

62 Illustration of instrument designs by Hooke, including (bottom left) his clock-driven equatorial quadrant. The top right-hand diagram shows how, by means of a 'Hooke joint', the instrument could be transformed into an altitude quadrant with a siderial motion. (Hooke, 1674)

once. The movable sight had a clamp and tangent screw, so that fine adjustment was by a micrometer screw, fitted with a graduated disc and pointer. (Hooke recommended that the device could be made by the clockmaker Thomas Tompion.) The instrument could be levelled, for taking altitudes, by the newly invented bubble-level (spirit-level). It was, alternatively, mounted equatorially and, most revolutionary of all, had a clockwork drive controlled by Hooke's invention of a conical pendulum clock. The bob of a conical pendulum moves in a circle, so it can be used to control a continuous driving force, suited to a telescope following the motion of the heavens, whereas an ordinary pendulum releases a force at discreet intervals. Once targeted, the quadrant would continue to track a star without further adjustment. Finally, the same driving force could be diverted, by means of a Hooke joint, into a vertical direction and used to drive the quadrant, in a vertical plane, so as to maintain the same celestial azimuth. This marvellous conception was never realized, but it is an indication of Hooke's mechanical vision.

Flamsteed and Halley were two other English proponents of telescopic sights, and both tried to convince Hevelius of their advantages, without the vigour – some would say tactlessness – of Hooke. In 1679 Halley visited Hevelius, at the instigation of the Royal Society, in an attempt to resolve the dispute. He spent some time observing with Hevelius and was surprised and impressed by the accuracy and consistency of his results. Neither party was converted, but they parted on good terms.

The Royal Society's efforts in the affair seem in general to have been directed towards preserving goodwill among the astronomers. Its Secretary, Henry Oldenburg (*c.*1618–77) wrote to Hevelius in 1677 to explain Flamsteed's position, and while he was justifiably confident that time would prove the value of telescopic sights, he reassured Hevelius that his observations were properly esteemed by the Society:

> Feeling that telescopes do excell the common plain sights for taking observations he [Flamsteed] does hope you will allow him to enjoy his own opinion, no less than do his friends at Paris, who as we hear use nothing but telescopic sights, the preferability of which demonstration and experience will fully prove, unless we very much mistake. And meanwhile I hope you will not think there is so much friendship and intimacy between Flamsteed and Hooke that the two are conspiring together to destroy the value of your observations.
>
> (MacPike, 1937, p.83.)

Oldenburg was informing Hevelius of what most of the local community must have known; the superiority of telescopic sights was one of the few matters on which Hooke and Flamsteed were able to agree, and a conspiracy between the two was scarcely credible. In 1674 Hooke was engaged on his magnificent quadrant design, but as his diary notes on 28 May, he failed to impress Flamsteed: 'I shewd Flamstead my quadrant. He is a conceited cocks comb' (Hooke, 1935, p.105). In 1677, the year of Oldenburg's letter, Hooke recorded on 27 November: 'At Tompions. Flamstead an Ignorant impudent Asse' (*ibid*, p.330).

Scientific societies and national observatories

We have seen Oldenburg cite the popularity of telescopic sights in Paris. This support was particularly important, for its source was a prestigious observatory, generously funded by Louis XIV; it was the other royal astronomical foundation of the later seventeenth century. The Observatoire de Paris and the Royal Observatory at Greenwich, through rivalry as much as anything, would in time generate the concept of a national observatory.

Both observatories were established soon after the foundations of scientific societies, and those societies were influential in both cases. The Royal Society had been founded in 1660 by a group meeting regularly at Gresham College. The Académie des Sciences began in 1667, and its astronomers used instruments in the Jardin de la Bibliothèque du Roy before the Observatory's foundation; these were a 9 ft quadrant and a 6 ft sextant, both with open sights. The foundation instruments of the Observatory, however, included fixed instruments with telescopic sights, such as two 6 ft mural quadrants. Picard's instruments also went to the Observatory at his death.

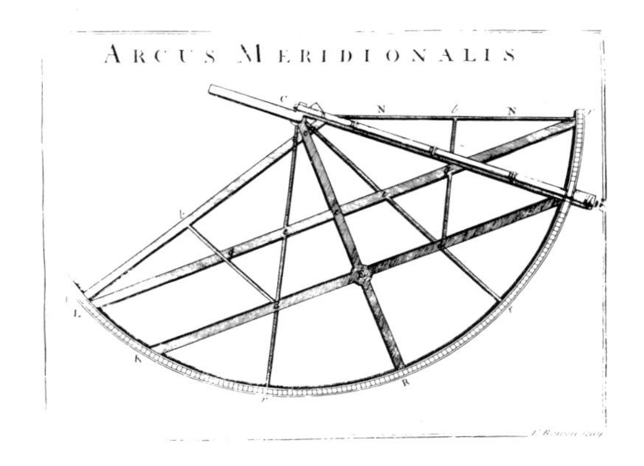

63 Flamsteed's mural arc of radius 7 ft, with a telescopic sight and counter-weight. It was divided by diagonals to 5 seconds of arc. (Flamsteed, 1725)

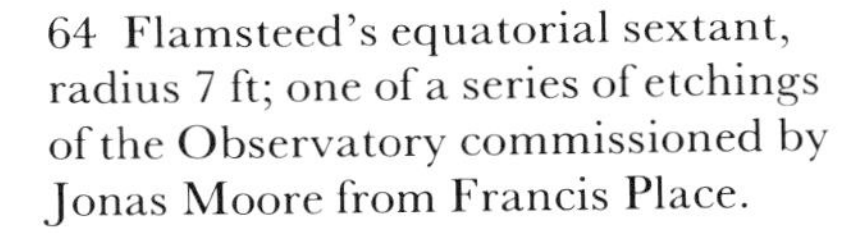

64 Flamsteed's equatorial sextant, radius 7 ft; one of a series of etchings of the Observatory commissioned by Jonas Moore from Francis Place.

Plans for the Observatoire de Paris were begun in 1667 and the building completed in 1672. The first director was the Italian astronomer Jean-Dominique Cassini (1625–1712), who at Louis XIV's invitation, came to Paris in 1669, already celebrated as a telescopic observer. Also resident in the Observatory from shortly after its completion was the Danish astronomer Ole Christensen Römer (1644–1710), who was introduced to Parisian science by Picard and worked in the Observatory until 1681. His observations of the satellites of Jupiter, aimed at a method of finding longitude, led him to conclude correctly that the transmission of light was not instantaneous. This was indicated by apparent discrepancies in his measurements related to a satellite's distance from Earth.

After his return to Denmark, Römer was appointed Astronomer Royal and given charge of the Copenhagen Observatory. He was subsequently responsible for profound developments in astronomical instrumentation. The transit instrument probably derived from an idea of Picard, but Römer had one built and put to use. The basic transit instrument consists of a telescopic sight, precisely mounted in the meridian by means of a horizontal axis, supported on either side. The telescope can move in altitude, and there may be a setting circle, but this altitude function is subsidiary to the measurement of azimuth, by timing the 'transits' of objects across the centre of the field of view. Thus the instrument serves to define accurately the observer's meridian, and the principal transit instrument defines the meridian of an observatory.

Römer also had a meridian circle, and correctly argued its advantages over the traditional quadrant, which in fact continued to be the standard altitude instrument in observatories until the late eighteenth century. The circle allowed more accurate scale division and, by taking a number of readings from different index arms, the elimination of error due to imprecise centring and the reduction of that due to inaccurate scale reading. Römer was also innovative in using microscopes to assist with reading scales.

While the famous Cassini presided at Paris, the scarcely known Flamsteed was also taking advantage of the new instrumentation in equipping the Observatory at Greenwich. There were serious financial restrictions, however, as Charles II had made no provision for fitting out his new observatory with instruments. Partly with the help of his patron Sir Jonas Moore, and partly at his own expense, Flamsteed managed to provide a 10 ft mural quadrant (designed by Hooke), which was replaced in 1689 by a 7 ft mural arc (fig. 63) of 140°. Each, of course, had a telescopic sight. He had also a 7 ft equatorial sextant (fig. 64); again Hooke was involved with the design, which bore some relation to his earlier scheme for an equatorial

65 Astronomical quadrant, English, early 18th century, radius 900mm, with fixed and moveable telescopic sights, the latter adjusted on the diagonal scale by a tangent screw. The two semicircles comprising the mount allow the quadrant to be set in any plane. Edinburgh, Royal Museum of Scotland.

quadrant. There were two telescopic sights, and the original division was by a worm or endless screw moving on the limb, with subdivision of the degrees by the revolution of the worm. Flamsteed afterwards added a conventional division with diagonals.

The foundation of national observatories, which were not dependent for their continued existence on the activities of individual astronomers, would have an important influence on the development of instrumentation. The observatories would commission instruments from successive generations of makers, as successive directors sought to up-date their equipment, in order to maintain a leading role for the observatory. An important public commission of this kind could establish a maker's reputation and was a significant commercial prize. The observatories would foster improved standards of accuracy, which would influence the entire trade, even if only the very best makers were directly involved.

The scientific societies also played important roles that were to be consolidated in the eighteenth century. The Académie was to commission instruments for a variety of projects and expeditions. The Royal Society would become a forum to which the very best English makers had access, and this promoted an informed commitment to the scientific enterprise. Both societies fostered communication and dissemination of ideas, through meetings, correspondence and the publication of journals.

In Chapter 6, some of the consequences of these changes in the social and institutional organization of science will become evident. But a more profound legacy of the seventeenth century was that of conceptual change: instruments were now central to the practice of both natural philosophy and mathematical science, and the distinction was no longer the great intellectual gulf it had been formerly. Further, there was enormous confidence that progress, intellectual and material, would inevitably flow from the scientific programme made possible by the new instruments. This confidence was, of course, most enthusiastically expressed by Hooke in his *Animadversions*:

> These [schemes for instruments] I mention, that I may excite the World to enquire a little farther into the improvement of Sciences, and not think that either they or their predecessors have attained the utmost perfections of any one part of knowledge, and to throw off that lazy and pernitious principle, of being contented to know as much as their Fathers, Grandfathers, or great Grandfathers ever did . . . Let us see what the improvement of Instruments can produce. (Hooke, 1674, p.45.)

5

The Early Specialist Trade

THOUGH THE CLEAREST HISTORICAL record has been left by the astronomers and mathematicians, we should not forget that the whole enterprise could not have been sustained without craftsmen, nor have progressed far without craft specialists. Since the specialist depends on a substratum of craft skills and resources – the availability of appropriate materials, the development of more basic metal-working techniques, and so on – sustained and successful instrument-making traditions are most likely to occur where a number of influences, both social and personal, intersect.

A healthy commercial community helps to support the existence of more technical skills, for even a specialist must generally look to relatively commonplace items for his regular income. Basic technical resources must also be present, or it may be that their availability will influence the specialist product, as with the ivory sundials of sixteenth-century Nuremberg or seventeenth-century Dieppe. These resources are more likely to be channelled towards practical mathematics where there is appropriate expansion in navigation, trade or colonization; in other words, more general economic and political factors enter the complex equation. The final catalyst is often an individual, or a small group or succession of makers, with innovative ideas, enthusiasm for mathematics and a vision of its practical potential.

Some early makers

When Regiomontanus settled in Nuremberg in 1471, this set of conditions was almost complete. He had chosen the town, he explained,

> . . . not only on account of the availability of instruments, particularly the astronomical instruments on which the entire science of the heavens is based, but also on account of the very great ease of all sorts of communication with learned men living everywhere, since this place is regarded as the centre of Europe because of the journeys of the merchants. (Rosen)

He thus appreciated the links between specialist instrument-making and general commercial and intellectual activity; Nuremberg was on the main trade-route between Italy and the Low Countries. It was nonetheless through his efforts that this potential was fully realized:

66 Simple theodolite by Jacobus de Steur, late 17th century, diameter 187mm. The four fixed slit-and-window sights allow the instrument to be used as a surveyor's cross, and by suspending it by the shackle on the left, it can also measure altitudes. Instruments of this type were commonly made in the Netherlands and are sometimes known as 'Dutch' or 'Holland' circles. Cambridge, Whipple Museum.

67 (*below*) Altazimuth theodolite by Erasmus Habermel, late 16th century, diameter (horizontal circle) 148mm. The vertical arc can be removed, probably to convert the instrument to a simple theodolite. The plumb-line for reading altitudes is missing. Oxford, Museum of the History of Science.

Nuremberg and nearby Augsburg, both famous for general metal work, were the most important centres of instrument-making throughout the sixteenth century.

Georg Hartmann (1489–1564) was a leading maker of precision instruments in this tradition, who settled in Nuremberg in 1518, and whose instruments included astrolabes, quadrants, armillary spheres and globes. Erasmus Habermel (d.1606) probably worked in Augsburg and nearby Regensburg before settling in Prague as instrument-maker to Emperor Rudolph II. He was in the emperor's service at the same time as Tycho Brahe and later Johannes Kepler. One of the most celebrated makers was Christopher Schissler (*c.*1531–1609) of Augsburg, who produced some of the finest instruments of the period.

As interest in practical mathematics spread, so new centres of instrument-making were founded. Important makers in France in the sixteenth century included Michael Coignet (1549–1623) and Philippe Danfrie, who designed and made surveying instruments. Under the influence of Gemma Frisius and Mercator, who was, of course, a maker in his own right, a school of instrument-making was developed in Louvain, led after Mercator's departure by Gemma's nephew Gualterus Arsenius. It is significant that one of the first commercial instrument-makers in England, Thomas Gemini (*c.*1510–62), was of Flemish origin and may have been associated with the workshop at Louvain.

Navigational and territorial ambitions, characteristic of the Elizabethan period, prepared England for the development of practical mathematics, which had active propagandists in Digges and Dee. By the late sixteenth century a native commercial trade in instruments was established, and the leading and most skilled maker was Humphrey Cole (*c.*1530–91). The range as well as the quality of Cole's surviving instruments is impressive, but even so he was also employed outside mathematics: he engraved plates for printers and was a sinker of stamps at the Mint.

Other makers in England in the period include Augustine Ryther (*fl.*1576–95), an engraver of maps and charts as well as a maker of instruments, and James Kynvyn (*fl.*1570–1610), who was recommended by Digges and Blagrave. Ryther's talented apprentice, Charles Whitwell (*fl.*1590–1611) was particularly responsible for carrying the reputation of the new English school of instrument-making into the seventeenth century, and the style of his work indicates a continuity in craft tradition with, on the one side Humphrey Cole, and on the other Whitwell's famous apprentice Elias Allen (*fl.*1606–54).

It became customary for mathematical authors to recommend makers who could supply the instruments they described, and an early instance appeared on the title-page of William Barlow's *Navigator's supply* of 1597: 'The instruments are made by Chas. Whitwell over against Essex House, maker of all sorts of mathematicall instruments and graver of all these portratures.' The links between engraving instruments and engraving plates for mathematical books were sustained also by Allen, probably the most accomplished maker of seventeenth-century England. He prepared plates for Gunter, and Gunter in turn recommended him for supplying metal instruments, and John Thompson (*fl.*1609–48) for similar work in wood. Rathborne gave his readers the same advice. Allen was also recommended by Oughtred, and supplied some of the instruments taken on the voyage of Thomas James. A great many of his instruments have survived, probably the largest being a 7 ft quadrant.

Mid-century mathematical instrument-making in England is represented at its best by Anthony Thompson (d.1665) and Henry Sutton (d.1665). Thompson succeeded to the Hozier Lane shop of John Thompson and was employed by Wren and by Hooke. Sutton was especially renowned for his skills as an engraver of instruments. He combined the expertise of copper-plate engraving with instrument-making, by printing instrument designs on paper, which could then be pasted onto wooden boards (fig. 74), to form a cheap substitute for a specially engraved instrument in brass.

Walter Hayes (*fl.*1651–92) was a skilled maker towards the end of

68 Graphometer by Michael Butterfield, *c.*1700, diameter 280mm, with a diagonal scale. The large central compass allows the instrument, unlike the earlier design by Danfrie (fig. 49), to be used as a circumferentor. Geneva, Musée d'Histoire des Sciences.

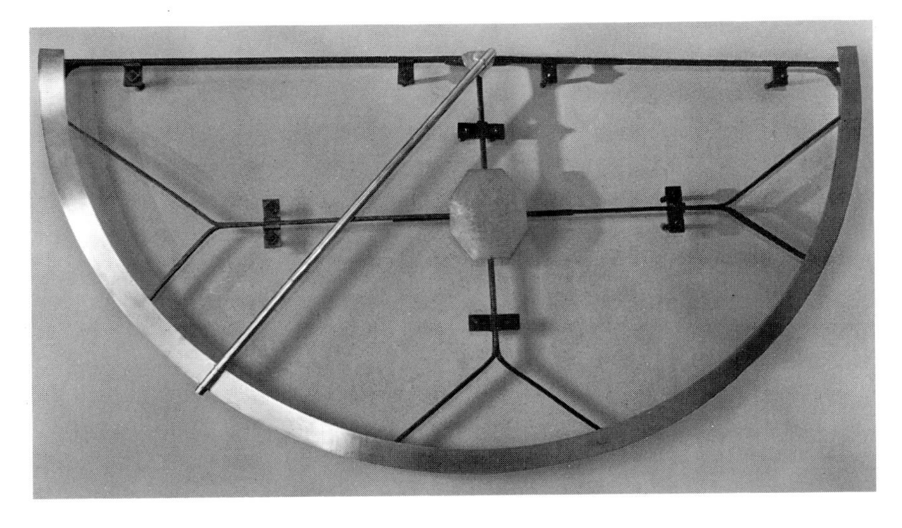

69 Mural semicircle by Lusverg, 1704, approximate radius 4ft, with a telescopic sight and divided by diagonals. The semicircle is an unusual astronomical instrument. The rather flimsy bracing can be contrasted with the later instruments of Graham and Sisson. Bologna, University Observatory.

the century, connected at one remove through the apprenticeship system to Elias Allen. One of his own apprentices, the famous Edmund Culpeper (1660–1738), succeeded to his business and to his trade sign, a pair of crossed daggers. By then there were many specialist mathematical instrument-makers in London, for the trade had grown vigorously during the century. Nonetheless, when Flamsteed came to commission instruments for the Royal Observatory, he did not turn to a craftsman generally regarded as an instrument-maker, but to the clockmaker Thomas Tompion (1638–1713). Hooke, who employed Tompion in making astronomical instruments, described him as 'A person deservedly famous for his excellent skill in making Watches and Clocks, and not less Curious and deserving in the construction and handworking of other nice mechanick Instruments' (Symonds, 1951, p.296). Rigid distinctions between engravers, clockmakers and instrument-makers are inappropriate to this period; one of Tompion's partners, the clockmaker George Graham (*c.*1674–1751), also became one of the most celebrated instrument-makers of the eighteenth century.

A few mathematical instrument-makers from early seventeenth-century France are known by signatures on instruments, but only with the founding and subsequent commissions of the Observatoire de Paris does much more specific information on makers emerge. Those who were principally opticians are excluded from this study, but among the mathematical makers Chapotot (*fl.*1680–88) was well known, having published accounts of his instruments and worked for Cassini and La Hire. Grosselin (*fl.*1671) made some of the first large instruments for the Observatory. It is an interesting early instance of specialism in the trade that two instruments signed by Grosselin are also engraved: 'Divisé par Sevin à Paris' (Daumas, 1972, p.75). In time, it would become a fairly common practice for makers to hand scale division over to a specialist. Pierre Sevin (*fl.*1665–83) himself is known by a number of surviving mathematical instruments.

Two noteworthy makers towards the end of the century are Michael Butterfield (1635–1724) and Nicholas Bion (1652–1733). Butterfield, an Englishman working in France from at least 1677, is best remembered for the design of the pocket horizontal sundial that bears his name, but he also made instruments for astronomy and surveying. Bion is best known for his publications on instruments, in particular his *Traité de la construction et des principaux usages des instrumens de mathématique*, which first appeared in 1709 and became a standard work in the eighteenth century.

In Italy, even more than in France, the most celebrated makers were opticians, but in the last quarter of the century Jacopo Lusverg made very fine surveying and astronomical instruments, at first in Modena and later in Rome. It should be stressed that the makers mentioned here are only the best known and represent a very small fraction of those recorded in biographical dictionaries, museum inventories, etcetera. Most, however, are recorded through more commonplace instruments than the grand pieces for observatories or wealthy patrons. It would, therefore, be appropriate in this chapter to deal with a few of the instruments on which the bulk of a maker's everyday trade depended, and particularly those most closely related to astronomy and navigation.

The nocturnal

Regarded by many as an instrument for the navigator, the nocturnal (fig. 70) was used for finding the time at night, and a skilled practitioner might expect an accuracy to within a quarter of an hour. The instrument was introduced in the sixteenth century and remained popular until at least the middle of the eighteenth. Examples fall very obviously into two groups. There are the smaller, finely engraved nocturnals, commonly French or Italian, usually in brass and often combined with a sundial or lunar volvelle (showing the phase of the Moon). The earlier instruments are of this form. Then there are the larger, wooden nocturnals, usually in boxwood, generally English and dating to between the mid-seventeenth and mid-eighteenth centuries. Of the two general types, the wooden ones are more like shipboard instruments, and this is confirmed by the fact that they almost always carry additional scales related to latitude finding and tide prediction.

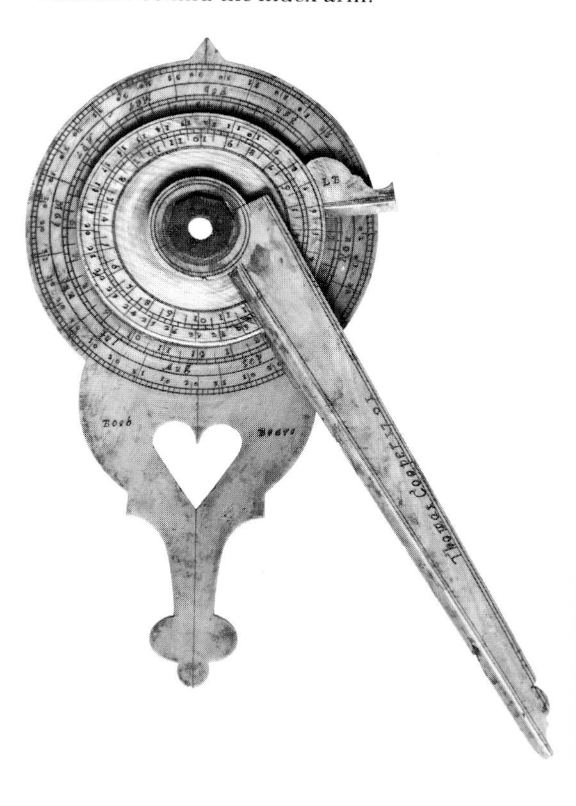

70 Boxwood nocturnal by Thomas Cooper, 1701, length 260mm, for 'Both Bears'. The 'LB' pointer, for orientating the time scale on the volvelle with respect to the date scale, is visible; the 'GB' pointer is obscured behind the index arm.

Because of the Earth's daily rotation, the stars appear to circle the pole once every twenty-four hours, and the orientation of the stellar pattern can be used to estimate the time. Strictly speaking, the apparent rotation is performed in a little less than twenty-four hours, because our ordinary timekeeping is governed by the Sun and not the stars and the Sun makes an apparent annual orbit through the heavens. The daily shift in the Sun's position relative to the stars represents the difference between the sidereal and solar days – the Earth must rotate a little further before the Sun appears once again to cross the observer's meridian – and this difference amounts to almost four minutes.

When finding time by a nocturnal, four minutes are neither here nor there, but the difference is, of course, cumulative throughout the year. The stars are in different positions in the sky at the same time on different dates. Thus any timekeeper based on stellar observation must allow for this, and a correction is built into the nocturnal: the time scale must first be adjusted according to the date.

71 The use of the nocturnal for finding the time at night. The instrument is held vertically at arm's length, while the observer sights the Pole Star through the central hole. The index arm is then aligned with two stars in the constellation of the Great Bear, and the hour scale, which has previously been set for the date, shows the time as 1.30 a.m. (Apianus, 1539)

The nocturnal consists of two circular plates (the larger one has a handle) and a long index arm, all pivoted together by a rivet through the centres of the plates and one end of the index arm. The rivet is pierced by a central hole. The larger plate has a scale of months and days, the smaller a scale of time, which is sometimes only half-divided into the twelve hours of night. The inner scale also has a pointer at twelve o'clock midnight, which the observer must first set to the date, thus adjusting the time scale for the time of year. The instrument is then held at arm's length (fig. 71), the Pole Star sighted through the central hole, and the index arm rotated until its bevelled edge just intersects the star for which the instrument was calibrated. This star is an index of the orientation of the heavens, and the time is simply read off the inner scale.

The 'index stars' commonly used were either the 'Pointers' in the constellation of the Great Bear (Ursa Major), which are in line with the Pole Star, or the bright star Kochab in the Little Bear (Ursa Minor). Often either index can be used, when the inner plate carries two pointers, indicated by 'GB' and 'LB', and such an instrument might be marked 'Both Bears'. Sometimes the inner plate has teeth at the hour divisions, so that by feeling and counting from the long midnight tooth, a rough time could be found in the dark.

This, the larger, form of nocturnal will also carry a time scale on the outer plate and a lunar scale, numbered 1 to 29, on the inner. These are used to find the time of high tide, given the Moon's age and the 'establishment' of the port in question. 'Establishment' – a constant for each port – is the time interval between the meridian transit of the full Moon or the new Moon (midnight and midday respectively) and the next high tide. The zero of the lunar scale is set to the establishment of the port on the time scale of the larger plate. The index arm is then set to the Moon's age on the lunar scale, and the time of high tide read off the time scale on the inner plate.

The reverse of the nocturnal has two concentric scales (if the instrument is for both bears), giving the corrections to the altitude of Polaris for use in finding latitude, and is an instrumental version of the 'Regiment of the North Star'. The scales are used in conjunction with the index arm, which is set to the star's position in the normal way.

A different type of nocturnal (fig. 72), described by Edmund Gunter as a 'Nucturnall for the use of seamen', is sometimes found on the reverse of an horary quadrant. It has a solid, circular volvelle,

72 Planispheric nocturnal on the reverse of a Gunter quadrant by Elias Allen, *c.*1630, radius 90mm. After orientating the planisphere (which matches the engraving in Gunter, 1623), the time scale is read at the appropriate date. Cambridge, Whipple Museum.

with an engraved planisphere marked with a few prominent stars and constellations, and a date scale. This moves over a time scale, divided to quarter hours. The planisphere is simply made to correspond with the orientation of the night sky, and the time is given against the appropriate date. When found, this nocturnal will usually be combined with our second commonplace instrument, Gunter's quadrant.

The Gunter quadrant

Edmund Gunter's was the most popular horary quadrant (fig 73) of the seventeenth and eighteenth centuries. It was also a portable astronomical instrument and calculator in the manner of the astrolabe. Gunter published an account as an appendix to his *De sectore et radio* of 1623, where he described his sector and cross-staff and a few other instruments.

The quadrant was based on the traditional form – fixed sights on one edge, a plumb-bob suspended at the apex, and a degree scale on the limb. On the face was engraved an outline projection of the sphere, from the equator to the limit of the Sun's declination (an arc coinciding with the limb and representing either of the tropics). To this was added a declination scale on one edge, and projections of the ecliptic and the horizon. The projection was for a single latitude, and was equivalent to one plate of an astrolabe.

Between the quarter circles representing the equator and the tropics were plotted lines of time (hour lines), according to solar declination (equivalent to date) and solar altitude. These lines were drawn in pairs to include both summer (Sun declines above the equator) and winter (Sun below the equator), and each line had to represent complementary hours before and after noon. By these expedients, the whole solar cycle could be contracted to a quadrant. Since solar altitude, declination and time are mutually dependent, one –

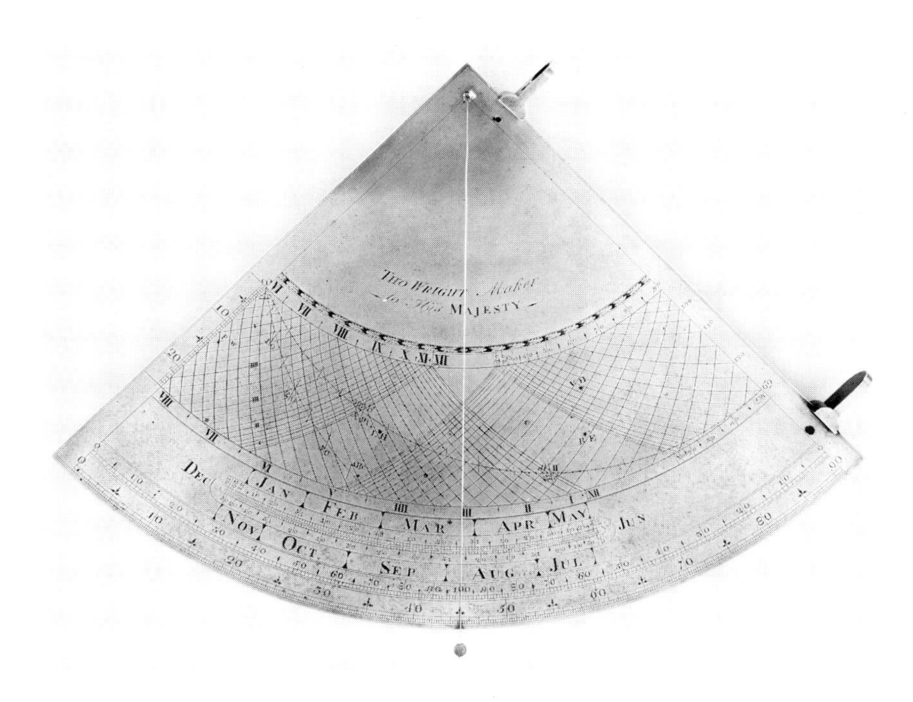

73 Gunter quadrant by Thomas Wright, *c.*1720. The plumb-line is present, but it lacks the bead, whose position on the thread is set by the declination scale on the left. London, Science Museum.

74 A paper horary quadrant by Henry Sutton, 1658, radius 277mm, pasted on to wood. This design of quadrant, known as the 'Sutton type', is less common than the Gunter projection. Cambridge, Whipple Museum.

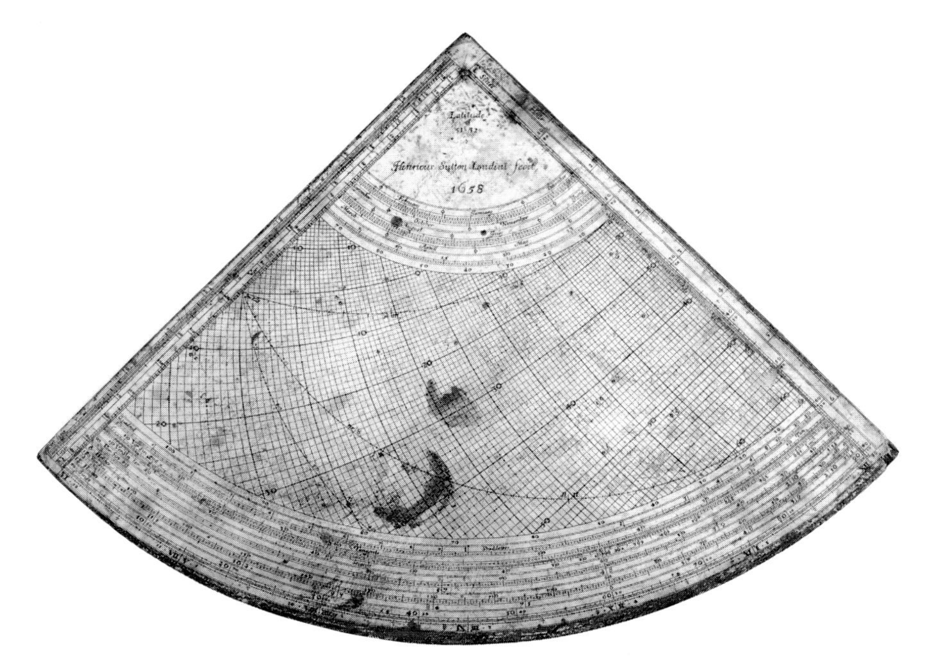

normally time – can be determined from the other two. To find time by the Gunter quadrant, a bead, which slides friction-tight on the plumb-line, is first set to the appropriate solar declination by the declination scale, and the Sun is sighted so that the plumb-line registers its altitude. The bead then indicates the time by the hour line on which it falls.

The ecliptic line permitted other calculations related to the solar cycle; the horizon line calculations of solar amplitude, times of the Sun's rising and setting and the length of the day. The inclusion of prominent stars on the planisphere allowed the time at night to be found from one of their altitudes. Space was also found for lines representing solar azimuths, plotted in pairs like the hour lines, according to solar declination and altitude. These allowed calculations relating solar altitudes and azimuths – useful in establishing the meridian, magnetic variation and so on.

Finally, Gunter included the traditional shadow square, and concluded his account with its customary applications to surveying. The whole was a marvellously ingenious device, and its ingenuity clearly appealed to contemporary interest in practical mathematics: examples are found engraved in brass, engraved in wood and printed on paper. It soon became a stock item for the mathematical instrument-maker.

The universal equinoctial ring dial

Our third stock item was, if anything, even more popular than the Gunter quadrant. It is a form of sundial and must serve to represent the many forms of sundial available from the makers. This one – the equinoctial ring dial – is an appropriate example because, in addition to its popularity and elegance, its design was associated with some of the leading mathematicians mentioned, and it links together the three disciplines of astronomy, navigation and surveying.

There are two forms. The early and uncommon one had three concentric rings and was first described by Gemma Frisius; it was known as the 'astronomer's rings' or 'astronomical ring dial' (fig. 9). The outer, 'meridian ring' had a latitude scale of degrees and a sliding suspension shackle to set the instrument for latitude. Thus the dial was 'universal' – it could be used in any latitude. Mounted at right angles to the meridian ring and at the zero of the latitude scale (thus parallel to the equator), was the 'hour ring' or circle of right ascension. Twelve o'clock is found at either pivot, representing the Sun crossing the meridian.

In the early form a third, innermost ring is mounted at 90° on the latitude scale – thus at right angles to the equatorial plane – and is equipped with a pair of sights. The whole then constitutes a portable and versatile astronomical instrument, and Gemma also illustrates its application to surveying.

More commonly (fig. 76), this inner component is a bridge, set across the outer ring, 90° distant from the axis of the inner. It has a longitudinal slot in which slides a small cursor with a central hole, and it carries two equivalent scales on either side, one marked with dates and the other with solar declination. The design of this form – the common equinoctial ring dial – is usually attributed to William Oughtred. It was often recommended to seamen, on account of its universal character.

In use, the suspension ring is first set for latitude and the cursor set to the date. The hour ring, pivoted within the meridian ring so that the dial folds flat for convenience, is folded out and the dial suspended freely. It is carefully turned until the spot of sunlight from the hole in the cursor falls on the hour ring, and indicates the time.

75 (*left*) Azimuth compass by Ralph Walker, *c.*1793, diameter 270mm. The hour band is adjusted for latitude on the semicircular scale and the pin-hole set in the solar declination slot. The dial is rotated on its azimuth mount, and is in the meridian when the spot of light falls on the hour scale. The variation can then be read off the card. Greenwich, National Maritime Museum.

76 (*right*) Universal equinoctial ring dial, 18th century, diameter 152mm. The date scale (equivalent to solar declination) can be seen on the central bridge, the hour scale on the inner ring. The outer ring shows the 'nautical ring' (the suspension ring is not in its working position) with the latitude scale for the sundial on the other side.

The dial's functioning is easily understood. The daily apparent motion of the Sun can be considered as a circle parallel to the equator, at an angular distance from it corresponding to the solar declination. The shadow of a gnomon or style parallel to the Earth's axis will therefore move uniformly on a circle parallel to the equator. This would be the case if the bridge were simply a gnomon, since the hour ring is indeed parallel to the equator. The divisions on the hour ring are thus equal and the dial described as 'equinoctial'.

The dial would work with a simple gnomon, but would need to be orientated by a magnetic compass, with the meridian ring due north and south. It is the sliding cursor that gives the dial the additional refinement of being self-orientating. With the cursor set at a point on the bridge corresponding to the Sun's declination, the spot of transmitted light will move around the hour ring during the day, if the meridian ring is properly orientated. Thus, as the dial is rotated at a given time, the spot will cross the hour ring only when the meridian ring is indeed in the meridian. As well as telling the time, if such a dial is mounted above a magnetic compass, it will measure the magnetic variation.

There is often found a further link with navigation. A 'nautical ring' or 'sea ring' was an instrument first described in the late sixteenth century. It was an alternative to the mariner's astrolabe for measuring solar altitude at noon to discover the latitude. A wide ring, suspended when in use, had a pin-hole to cast a spot of light on to a 90–degree scale engraved within the ring. Its advantage over the astrolabe was that the scale was magnified; in fact it extended across half the circumference of the ring.

An English universal equinoctial ring dial will, more often than not, have a nautical ring engraved on the reverse of the meridian ring. The dial is then used flat, the suspension shackle set at the zero of the latitude scale, and a pin gnomon placed in the hole provided, so as to cast a shadow on the degree scale.

As we reach the close of our first period division – from ancient to early modern times – the universal equinoctial ring dial is an elegant reminder of the conceptual unity of the central instruments of astronomy, navigation and surveying, epitomized in the notion of the divided circle. This unity was nowhere better expressed than in the first English astronomy textbook, Robert Recorde's *Castle of knowledge* of 1556:

> Althovghe there be many and wonderfull instrumentes wittely deuised for practice in Astronomy, as the Astrolabe, the Plaine sphere, the Saphey, the Quadrante of diuerse sortes, the Chylynder, Ptolome his rules, Hipparchus rules, Tunsteedes rules, The Albion, the Torquete, the Astronomers staffe, the Astronomers ringe, the Astronomers shippe, and a greate numbre more, whiche hereafter in tyme you may knowe, yet all these are but parts, or (at the most) diuers representations of the Sphere.
>
> (Recorde, 1556, p.35)

6

The Heroic Age

HE EIGHTEENTH CENTURY REPRESENTS the heroic age of the individual craftsman. This is not to say that each maker was wholly responsible for every instrument that bears his name – far from it. Large workshops were established – particularly later in the century – employing assistants and apprentices. Individual parts, such as the optical components of mathematical instruments, were bought in from specialists. A scale might be divided in a workshop other than that of the instrument's principal maker. Instruments were engraved with the names of makers who had no other contact with their manufacture, and who acted in such instances only as retailers.

Yet there is no doubt that a maker's abilities as a mechanic were vital to the success of his workshop, large or small. It was an age of change and expansion in instrumentation, and a maker needed a sympathetic understanding of materials and mechanical techniques, to appreciate how they might be moulded to new needs or when they should be replaced. Other personal qualities were important, such as innovative ability and a familiarity with contemporary scientific interests and goals.

The rewards were considerable, for mechanics could acquire unprecedented status within the national scientific community and, in the most outstanding examples, an international reputation. National respect was especially characteristic of England, which occupied a commanding position by the later eighteenth century. One of the aims of this chapter is to explain how this came about.

Instrument-making on the Continent

England and France represent the most important centres of mathematical instrument-making in the eighteenth century. Though the Netherlands were important for the manufacture of instruments of natural philosophy, there were only a few significant makers on the mathematical side. In navigation, G. Hulst van Keulen and J. M. Kleman, both of Amsterdam, were active at the latter end of the century.

In Germany, too, there is little to record until the second half of the century, though sundials continued to be manufactured in

77 Two late 18th-century surveying compasses by Adams (left), diameter 130mm, and by Lennel, diameter 128mm, with different types of sights. Both fold into flat boxes, one with a hinged, the other with a sliding, lid. Cambridge, Whipple Museum.

78 (*opposite*) Graphometer with two telescopes by Lenoir, late 18th century, diameter 198mm. Although there is a telescopic sight for aligning the azimuth index, there is no vertical arc (cf. fig. 101). The lower telescope is used to establish and check the orientation of the instrument (equivalent to Danfrie's 'alidade des stations', fig. 49); it is more often included on continental instruments, including theodolites, than on English ones. Cambridge, Whipple Museum.

Nuremberg and Augsburg. The foundation of the Breithaupt workshop, however, in Kassel in 1762 is noteworthy, headed first by Johann Christian Breithaupt (1736–99) and maintained by his sons Friedrich Wilhelm and Carl Wilhelm. The largest recorded instrument by J. C. Breithaupt seems to have been a 6 ft mural quadrant for the Kassel Observatory.

Georg Friedrich Brander (1713–83) of Augsburg made an impressive range of very fine instruments, including a number to his own designs, which were described in a series of published pamphlets. From 1775 he was in partnership with his son-in-law Christian Caspar Höschel (1744–1820). Within our field, Brander's surveying instruments are most numerous and distinctive, but he also made very considerable instruments of astronomy. These included a series of instruments for the Jesuit Observatory at Ingolstadt: a 6 ft sextant, an 8 ft mural quadrant and a 12 ft equatorial sector.

In Italy, Domenico Lusverg carried the fine tradition of his father's workshop into the eighteenth century, but in general, though there were able opticians, there were few subsequent mechanics of note. Italian astronomers generally bought their large measuring instruments from makers in England and France. In France, there were a number of accomplished mathematical instrument-makers, and the century can be thought of in terms of two epochs, associated with two leading figures – the early part with Langlois and the successors to his workshop, the later with Lenoir.

Jacques Lemaire (*fl.*1720–40) is generally associated with the manufacture of sundials, and his son Pierre (*fl.*1739–60) also made surveying and navigational instruments. However, the workshop founded by Claude Langlois (*fl.*1730–51) dominated the competition for important commissions from about 1730, and Langlois himself was given an official position as mechanical instrument-maker to the Académie des Sciences. With his office went accommodation in the Louvre, so that his signature can be found embellished: 'Ingénieur Aux Galleries du Louvre'.

Langlois benefited enormously by his official connection with the

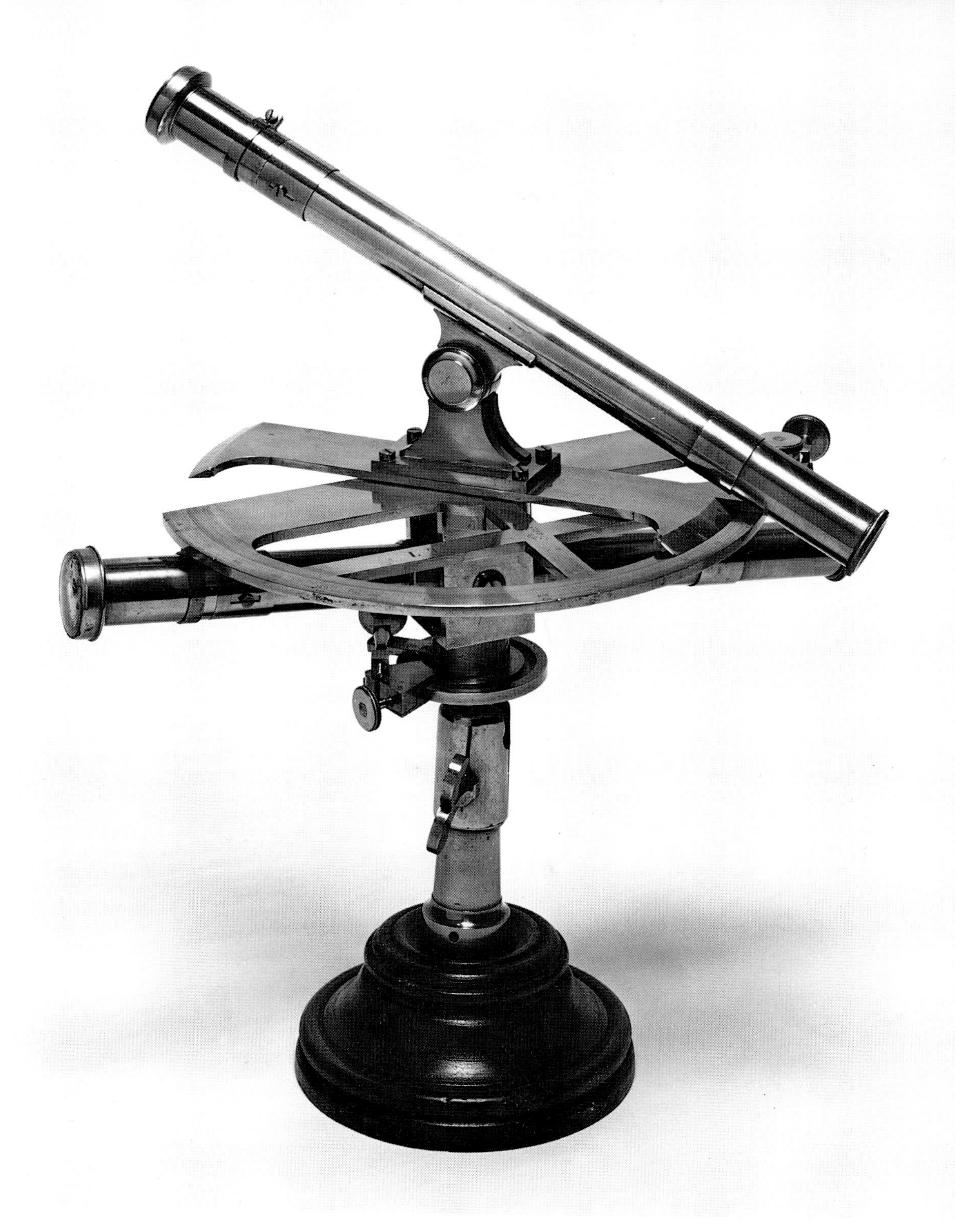

Academy and the commissions offered to him by the Paris Observatory. His large instruments for the Observatory were two 6 ft mural quadrants and a 6 ft sector. He gained greater fame, however, from his work for the Academy, probably because it derived from their most celebrated enterprise of the eighteenth century.

In the tradition of the geodetic work initiated in France by J. D. Cassini, the Academy mounted two important expeditions in 1735. One was to Peru, under the leadership of Charles-Marie de La Con-

79 Two plain theodolites with 'inverted' arcs, one by Adams (left), c.1780, diameter 120mm (horizontal circle), the other by Heath & Wing, *c.*1750, diameter 142mm. The vertical arc on the earlier instrument has rack-and-pinion motion and, in addition to a degree scale, a gradient scale and one of 'Links Diff. Hypo & Base', which gives the number of links to be subtracted from each chain length measured along the ground to find the horizontal distance. The later instrument has the telescope in Y bearings. Cambridge, Whipple Museum.

damine (1701–74); the other, led by P. L. Moreau de Maupertuis (1698–1759) went to Lapland. The object in each case was to measure the linear distance covered along a meridian by a degree of latitude. The results were intended to settle the contemporary controversy over the precise figure of the Earth. Newtonian gravitational theory, in opposition to the Cartesian natural philosophy still dominant in France, predicted a flattening of the Earth and a 'shorter' degree towards the poles. Precision instruments for measuring latitude (quadrants or zenith sectors) and for performing triangulations would be crucial to the success of such a project.

The whole enterprise represented a significant example of national patronage in mathematical science, and an archetypal instance of the scientific expedition, equipped with the finest instrumentation, specially commissioned for the occasion. Here was a precedent that the English, as well as the French themselves, would follow later in the century.

While some of the expedition's instruments were obtained from George Graham, others – notably the portable quadrants – were by Langlois, and he continued to supply instruments for further French geodetical projects. C.-F. Cassini de Thury (1714–84, grandson of J. D. Cassini) and Nicolas-Louis de Lacaille (1713–62) used instruments by Langlois – a 2 ft quadrant and a 6 ft sector – for further geodetical work in France, and the equipment taken by Lacaille on his expedition to the Cape of Good Hope in 1750 included a 3 ft quadrant and a 6 ft sextant by Langlois. Prominent among Lacaille's work at the Cape were further measurement of a degree (the question was not yet positively settled) and positional observations of southern stars.

Langlois's successor – to his workshop and to his position as the Academy's engineer – was his nephew Canivet (d.1774). His output was not unlike that of Langlois – portable quadrants for astronomical and geodetical work, the occasional large observatory instrument, and everyday surveying instruments such as graphometers. A major commission was for the Jesuit Observatory in Milan: a 4 ft transit instrument, a 6 ft sextant and a 6 ft mural quadrant were all supplied during the 1760s. Successor to Canivet – to the workshop, but not to the Academy appointment – was Lennel (*fl.*1774–83). The quadrant was now nearing the end of its career as the principal portable instrument of astronomy, to be replaced in the first instance by the repeating circle, but the important instruments among Lennel's *œuvre* were quadrants and transit instruments. Though a very able maker, he was not an innovator.

If French instrument-making was at a low ebb during the third quarter of the century, a revival was signalled by the arrival of a maker with creative as well as mechanical ability. Like that of Langlois, the career of Etienne Lenoir (1744–1832) was forwarded by official commissions, such as supplying instruments for geodetical and naval expeditions – notably the instruments used to establish the length of the metre. The metre was based on a measurement of a degree of latitude, construed as a 'natural' standard of length. The

80 (*left*) Portable transit instrument by Lenoir, late 18th century, telescope length 535mm, with a ring base and a broad vertical semicircle, read by a central vernier to a minute of arc.

81 (*right*) Y level by Cary, early 19th century, illustrating the arrangement of four levelling screws between parallel plates set above the tripod.

82 A dual-purpose surveying instrument by Benjamin Cole, early 19th century, diameter 133mm. This instrument served as either a simple theodolite or as a circumferentor, according to whether the outer scale or the inner compass scale was consulted.

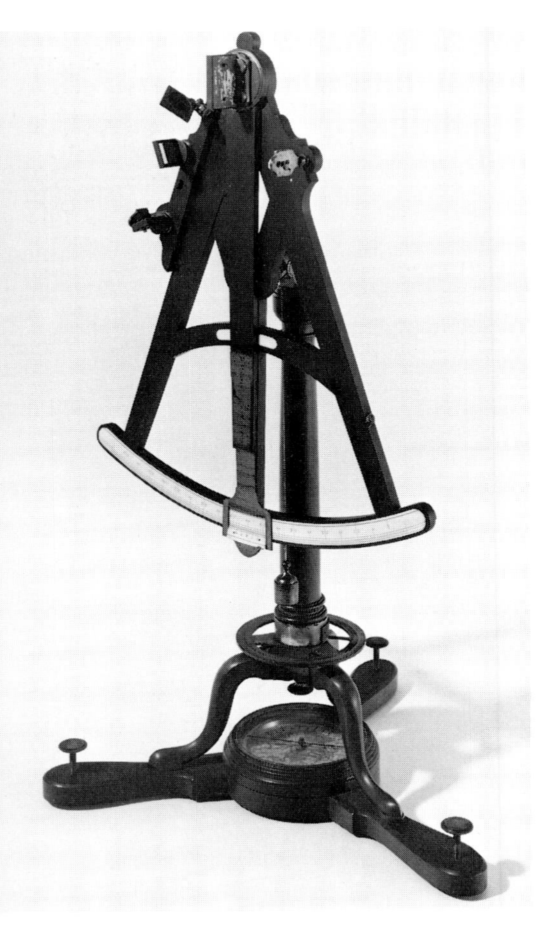

Lenoir workshop was able to handle major observatory instruments, and to capitalize on the important French invention of the reflecting circle, applied across our three disciplines of astronomy, navigation and surveying. By the time the workshop came under the charge of Lenoir's son Paul-Etienne (1776–1827) early in the nineteenth century, it was the inspiration for new confidence among the French that their instrument-making capacity was properly re-established and could hope to compete with the dominant centre in London.

It is clear that commissions from official bodies in France – notably the Academy and the Paris Observatory – were vital to the development of the trade. Only the leading makers could benefit directly from such patronage – not only in immediate financial terms but also by establishing a reputation – but their standards influenced the whole trade, for their workshops also turned out a wide range of more commonplace instruments. Two further examples of state patronage are worth noting, but have been left until now because they were international in character and involved in particular the London makers, as the principal rivals to those in Paris. The observations of the transits of Venus in 1761 and 1769, and the organization of a national survey were two respects in which the grand, officially sponsored projects, characteristic of science in eighteenth-century France, influenced activity in England.

Instrument-making in England

Any survey of the observatory instruments in use on the Continent towards the end of the eighteenth century will soon come upon the names of such English makers as Bird, Sisson, Dollond and Ramsden. We have already noted an early instance of an English export in the principal instrument taken on Maupertuis's expedition to Lapland – a zenith sector by George Graham. There is no doubt that London makers dominated the trade in mathematical instruments in the second half of the century. This is clear from the historical record, but was also well understood at the time.

If we look for examples of large fixed instruments at observatories in France, Italy and Germany, we know of mural quadrants by Sisson in Paris, Bologna, Pisa and Berlin; by Bird in Paris (two examples), Göttingen, Mannheim and Berlin; and by Ramsden in Milan and Padua. There were transit instruments by Sisson in Bologna and Florence; by Ramsden in Paris, Mannheim, Gotha, Leipzig and Palermo. There was a zenith sector by Sisson in Florence, and equatorial sectors by Sisson in Milan and Naples, and by Dollond in Kassel. There were altazimuth circles by Sisson in Naples and by Ramsden in Palermo.

This list is not exhaustive, but can easily be extended by looking further afield. In Amsterdam there was a meridian circle by Sisson, and mural quadrants by Sisson found their way to Gdansk (Poland) and Eger (Hungary). Eger also had a transit instrument by Dollond, as had Madrid, and transit instruments by Bird were in Madrid and Stockholm. In Uppsala there was a zenith sector by Sisson, and in Cadiz a mural quadrant by Bird. St. Petersburg was equipped by

84 Waywiser by W. Harris & Son, late 18th century, an instrument typical of waywisers signed by many different makers, but perhaps made by only a few. London, Science Museum.

Bird with a mural quadrant and a transit instrument, and Vilna (Lithuania) by Ramsden with a mural quadrant, a transit instrument and a meridian circle. Needless to say, British observatories in the eighteenth century were supplied from London.

Eighteenth-century London produced a series of outstanding makers of mathematical instruments: George Graham (*c.*1674–1751), Jonathan (*c.*1695–1747) and Jeremiah Sisson (*fl.*1736–88), John Bird (1709–76), Jesse Ramsden (1731–1800) and John (d.1807) and Edward Troughton (*c.*1753–1835). The Dollond workshop (John, 1706–61, and Peter Dollond, 1730–1820) is famous for optical work, particularly refracting telescopes, but also built major astronomical instruments. James Short (1710–68) was more exclusively an optician, specializing in reflectors, but their quality made them popular among professional as well as amateur astronomers, and with the addition of a micrometer (fig. 126) invented by John Dollond, they could be used for certain measurements in astronomy. Edward Nairne (1726–1806) also made fine mathematical instruments, mostly portable ones, but in addition he worked extensively on instruments of natural philosophy. As well as working on major commissions, most of these men were manufacturing everyday instruments: Sisson in surveying, Bird and Nairne in navigation, Ramsden, Dollond and Troughton in all disciplines.

Graham's influence was central to the establishment of the vigorous London tradition. He was a link with Tompion, and thus with the foundation instruments of the Royal Observatory. He was responsible for such innovative contributions as the dead-beat escapement and the mercury (temperature-compensated) pendulum, which together created the 'astronomical regulator' – a longcase clock that was also a measuring instrument in practical astronomy. He also devised the forms of mural quadrant and zenith sector that became standard in the eighteenth century. Known as 'honest' George Graham, he helped and advised Jonathan Sisson, John Bird, James Short and even lent money to the difficult mechanical genius John Harrison (*c.*1693–1776).

85 Waywiser dial, 18th century, 127mm square, divided into miles, furlongs, poles and yards. The miles and furlongs hand is shown, but a second hand for poles and yards is missing.

83 (*opposite*) Octant by Benjamin Martin, *c.*1760, unusual in being mounted on a wooden stand, as though for use on land in astronomy or surveying. There is an altazimuth mount, a pillar and tripod stand, and a tribrach base with three levelling screws and a compass. The plumb-line – useless at sea – could be used for levelling in the absence of a horizon. London, Science Museum.

86 Plotting protractor by Adams, late 18th century, diameter 203mm. A typical instrument used by the surveyor at the drawing-board, with a 360° divided circle and rack-and-pinion adjustment to two spring-loaded pricking pins.

We have seen that in France official bodies were important in the careers of the leading mathematical instrument-makers. The same was true in England, and here three organizations often feature in the career structures of the makers cited above. These were the Board of Longitude, the Royal Observatory and the Royal Society.

The Board of Longitude was founded in 1714 as an official response to the urgent need of a solution to the longitude problem. The problem was already well known, but public awareness of it had been heightened by the wreck on the Scilly Isles of four ships under the command of Admiral Sir Clowdisley Shovel, with the loss of nearly 2,000 men, and by the lobbying campaign of William Whiston (1675–1714) and Humphrey Ditton (1675–1714) on behalf of a fanciful scheme of their own. A fabulous prize was offered for a solution – £20, 000 for one accurate to within half a degree or 30 miles – and a committee was established to examine proposals. Thus the Board of Longitude, whose membership included the Astronomer Royal, the President of the Royal Society and mathematical professors from Oxford and Cambridge, was an official body specifically concerned with a question central to the practical mathematical sciences.

This official recognition, reinforced by the attraction of the prize itself, had a significant effect on precision instrument-making in England. Part of the Board's brief was to encourage relevant developments, even if they might – for the time being – fall short of a full solution, and there were lesser sums to be disbursed to this end. Two of the most important treatises on dividing mathematical instruments were published by the Board for the general benefit of the trade. These were John Bird's *Method of dividing astronomical instruments* (London, 1767) and Jesse Ramsden's *Description of an engine for dividing mathematical instruments* (London, 1777). Bird was recognized as the most skilled divider of his day and Ramsden's introduction of a dividing machine was to revolutionize the manufacture of portable

87 Marine compass by Adams, *c.*1776, 260mm square, with a fleur-de-lys north and decorated east points. An ordinary compass of the period, but with a famous signature. Greenwich, National Maritime Museum.

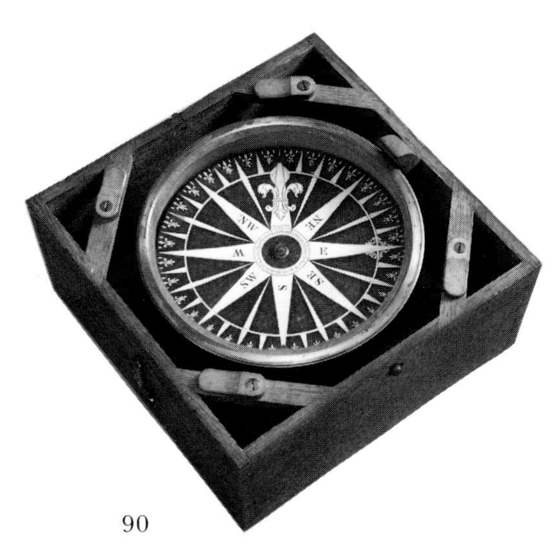

instruments. Both men received rewards from the Board. The Board could also commission trial instruments, and there was no more important commission than that which engaged Bird to construct reflecting circles for trial at sea, and which, as we shall see, led to the first marine sextants.

The Royal Observatory, too, had been founded with the aim of providing a longitude solution, and this very clearly influenced the kind of astronomy it undertook. Precision measurement was given the first priority and so it was vital that the Observatory was equipped with the most up-to-date instrumentation. During the eighteenth century, successive Astronomers Royal improved the stock of instruments, and a major commission by the Royal Observatory was a considerable asset to a maker's reputation. Sisson, Graham, Bird, Troughton and Peter Dollond all attracted such commissions, as did Ramsden, though he did not deliver, and, to a lesser extent, Nairne.

Perhaps the most distinctive feature of science in England, so far as the makers were concerned, was the role played during this period by the premier scientific organization, the Royal Society. Fellowship

88 Jesse Ramsden, holding a rule and dividers to symbolize his trade, but with his arm on the instrument that revolutionized it, his dividing engine. In the background is one of his most famous instruments, the Palermo altazimuth circle.

of the Society was open to the most able makers, and was granted to Graham, Ramsden, Edward Troughton, Short, John Dollond and Nairne. Each of them, and Peter Dollond also, published papers in the Society's *Philosophical transactions*. Most remarkably of all, the Society's highest award, the Copley Medal, was granted individually to John Dollond, to Ramsden and to Edward Troughton. This recognition by, and close links with, the scientific community gave the best English makers a commitment to the scientific enterprise, as well as a familiarity with its current interests and goals. It is interesting to contrast the situation at the Académie des Sciences, which was comparatively well endowed with funds and which employed mechanics in officially designated capacities. The Royal Society, on the other hand, had considerably less financial patronage to dispense, but admitted the best mechanics to fellowship.

Factors within the scientific community were important, but the more general economic and social conditions noted in other environments favourable to successful instrument-making should not be disregarded. Increased trade and wealth had created demand for the professional instruments of the navigator and surveyor. More widespread affluence and leisure had led to a fashionable interest in science as a hobby or entertainment. This was manifested in popular public lecture courses and textbooks, as well as in a new range of instruments aimed at the interested amateur, such as microscopes, telescopes, orreries, air-pumps, and electrical machines. The makers who catered primarily for this side of the trade, such as Benjamin Martin (1704–82) or George Adams, father (1704–72) and son (1750–95), though they attracted fashionable and even royal patron-

89 Plane table compass by Adams, mid-18th century, diameter 137mm. Typical of the period, the compass sits in an octagonal frame, with an extension to fit a slot beneath the table. The printed card is marked with the points of the nautical compass, divided to degrees and numbered 0-360 in both directions. Cambridge, Whipple Museum.

90 Plane table alidade by Fablmer of Strasbourg, *c.*1725, length 437mm. An early example of an alidade equipped with a vertical arc and telescopic sight. The rule has a linear scale and the arc is divided to ½° Cambridge, Whipple Museum.

age, were not admitted in the same way as the mathematical makers to the scientific community. They were not elected to the Royal Society.

International enterprise

Expeditions to observe the transits of Venus and large-scale, government-sponsored surveys were two notable occasions of trans-national influence in the second half of the century. In both cases, the more centrally organized and better-funded scientific community in France took the lead, while the technical resources of the London mechanical tradition allowed the English to emulate the French.

A transit of Venus is a rare astronomical event. It is the intervention of Venus between the Sun and the Earth, when the planet is seen as a tiny dark disc on the face of the Sun. Such transits occur in pairs, the two members being separated by eight years and the pairs themselves by over a hundred years. Eighteenth- and nineteenth-century interest in them derived from the fact that they allow a determination of the distance between the Earth and the Sun (the 'astronomical unit'), which in turn gives a scale to the entire solar system. The technique is a form of triangulation, with the planet's transit being observed and timed from widely separated stations on Earth.

Only one transit of Venus, in 1639, had ever been observed, and that by only two observers: imperfectly by William Crabtree (1610–*c.*1644) and completely by Jeremiah Horrox (1618–41), both observing in the north-west of England, and both friends of William

91 Altazimuth theodolite by Brander, 1760, diameter (horizontal circle) 122mm, both circles with endless screw motions. Munich, Deutsches Museum.

Gascoigne, the inventor of the eyepiece micrometer. The next transit was due in 1761 and the French were foremost in organizing expeditions, whose destinations included Siberia and the island of Rodrigues in the Indian Ocean. The English added expeditions to the Cape of Good Hope and to St Helena, while many other stations were set up elsewhere.

The main instruments required were telescopes fitted with micrometers and portable clocks, but others were taken along, such as quadrants and zenith sectors for determining latitude and longitude, regulating the clocks, and even for attempts to measure stellar parallax. Makers who received orders on the English side included Short, Dollond, Sisson and Bird. In 1769 the French went to Pondicherry in India and to California; British observers were found in Scandinavia and Hudson's Bay, while the most famous expedition was that led by James Cook (1728–79) in the *Endeavour* to Tahiti. The same circle of makers benefited from Royal Society orders.

The scale of activity was remarkable, in relation to the previous meagre history of such observations, and is some measure of the importance granted the practical mathematical sciences. H. Woolf's history of the episode cites 120 separate observations of the 1761 transit and 150 for that of 1769. English makers were best placed to take advantage of the sudden demand for instruments. For the 1769 transit the Imperial Academy of Sciences at St Petersburg placed orders in London for four quadrants and twenty-one telescopes, many with micrometer attachments. Even the French included English instruments in their equipment for 1769, with two Dollond telescopes going to California.

Dollond's development of the achromatic object glass gave London a very important edge in attracting commissions. With simple lenses, some dispersion of light into its component colours inevitably accompanies its refraction, so that the image is less precisely focused and has coloured fringes. Dollond corrected this 'chromatic aberration' by using lenses compounded of elements of different glasses – crown and flint – having different refractive properties. While the primary improvement was to observing telescopes, measuring instruments also benefited, since achromatic lenses allowed greater precision in aligning a sight or using a micrometer. While Dollond sought to preserve his patent, other makers, such as Bird, bought lenses from him, and Ramsden obtained a share in the patent in 1765 by marrying Peter Dollond's sister.

While France pursued a national survey in the mid-eighteenth century, under the promotion of Cassini de Thury and based on the meridian survey begun by his grandfather, there was no similar movement in England. Here a very considerable amount of private surveying was undertaken in the second half of the century and fine county maps were produced, but nothing was attempted on a national scale. The project that came closest to being a national one was a military survey of Scotland between 1747 and 1755 in the aftermath of the rebellion of 1745. Prominent in this enterprise was William Roy (1726–90), who was elected to fellowship of the Royal Society in 1767, and who promoted the idea of a national survey. His

own instruments were purchased from Sisson, Short, Dollond and Ramsden.

In 1783 Cassini de Thury proposed that the observatories of Paris and Greenwich should be linked by a triangulation across the Channel. There was a standing disagreement over the relative positions of the two national meridians. Roy saw this as an opportunity to raise again his project for a survey on a grand scale, and, with national pride at stake, the Royal Society arranged for the English side of the link to be executed by Roy. The French side had, of course, already been completed.

92 Plain theodolite by Francis Watkins, *c.*1770, diameter 110mm. An early example of the plain theodolite, having the telescope in Y bearings and a vertical semicircle to one side, and both motions by rack-and-pinion. The plotting compass is an original accessory. Cambridge, Whipple Museum.

93 (*opposite*) Repeating circle by Edward Troughton, *c*.1810, diameter (main circle) 480mm. This is the original form of repeating circle, used in astronomical instruments, with two pivoted telescopes and no mirror. The circle, which can be rotated into any plane, is divided on silver to 10 minutes of arc, and read by verniers on each of the four index arms to 10 seconds. The scale of the horizontal circle is gold. Cambridge, Whipple Museum.

Roy began by establishing an accurate base-line on Hounslow Heath in 1784. This operation was aided by two 100 ft chains from Ramsden and attracted much scientific attention. Even the king visited the site on 21 August and, it was thought, seemed satisfied with what he had seen. Roy was awarded the Copley Medal for this part of the work. The subsequent triangulation had to await Ramsden's completion of a great theodolite (fig. 142), whose horizontal circle had a diameter of 3 ft, and this was not delivered until July 1787. The whole survey was completed the following year.

Ramsden subsequently received an order for a similar pair of chains and a 3 ft theodolite (fig. 143) from the East India Company for triangulation in India. The Company refused to pay Ramsden's revised price and the theodolite was purchased by the Board of Ordnance for £373 14s in 1791. In this year, the Board resumed Roy's work (he had died in 1790), by initiating what became a national triangulation survey, which brought new commissions for Ramsden, including an 18 in theodolite and a zenith sector.

This activity promoted a recognition of the importance of precise survey and accurate maps, and raised the standards expected of the finest instruments. The Board of Ordnance's new responsibilities ensured that, as its work extended, more commonplace instruments would also improve. Surveying in Britain now joined astronomy and navigation in having an official body concerned with its development, with maintaining and improving standards of accuracy, and with dispensing commissions for the best instruments.

If the seventeenth century was characterized by conceptual innovation, the eighteenth was notable for the creation of organized structures – particularly at the official or semi-official level – relevant to the promotion of practical mathematical science, and for considerable expansion in demand for instruments. The period between the late seventeenth century and the early nineteenth saw the beginnings of national observatories, of national scientific societies, and of bodies concerned with the development of navigation and surveying. What had scarcely changed was the organization of the trade itself. Large workshops were just beginning to form; Ramsden's second workshop, in Piccadilly, was celebrated for employing over fifty men. But the old master and apprentices model was still the general pattern of working practice, and the trade had yet to come to terms with the idea of industrialization.

94 (*left*) Surveying quadrant by Christopher Schissler, 1579, 395mm square, with sights and scales for both measurement and calculation. The border is decorated in relief with scenes of surveyers in action. Oxford, Museum of the History of Science.

95 (*below*) Simple theodolite by Erasmus Habermel, *c.*1600, diameter 250mm, with a compass moving with the alidade and a prominent shadow square. Florence, Istituto e Museo di Storia della Scienza.

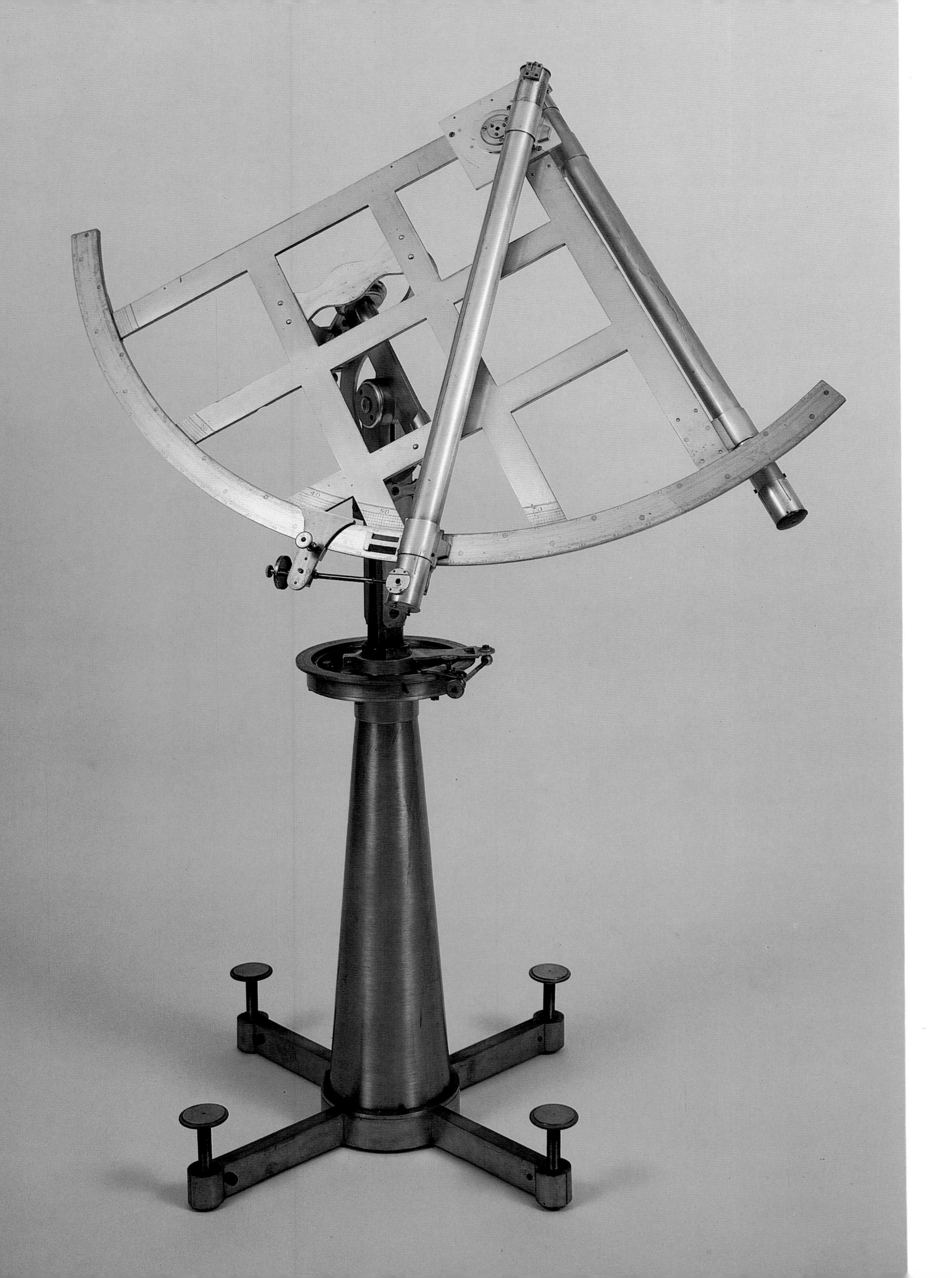

96 (*left*) Portable quadrant by John Bird, *c.*1765, radius 460mm., with a brass lattice frame, moveable into any plane, and fixed and pivoted sights. This instrument was used in the observatory of St John's College, Cambridge, and originally had a transversal scale. However Bird was engaged to modify his own quadrant by fitting a new limb, divided twice – to 90° and to 96 parts – and read by two verniers. Cambridge, Whipple Museum.

97 (*above*) Azimuth compass by Joseph da Costa Miranda of Lisbon, 1711, 228 by 228mm. The sights are vertical threads viewed through glazed apertures. They are used to measure the solar amplitude (the Sun's azimuth at sunrise or sunset) by the compass, and the card had raised divided edges east and west to facilitate the measurement, which is compared with the true amplitude found from tables. Cambridge, Whipple Museum.

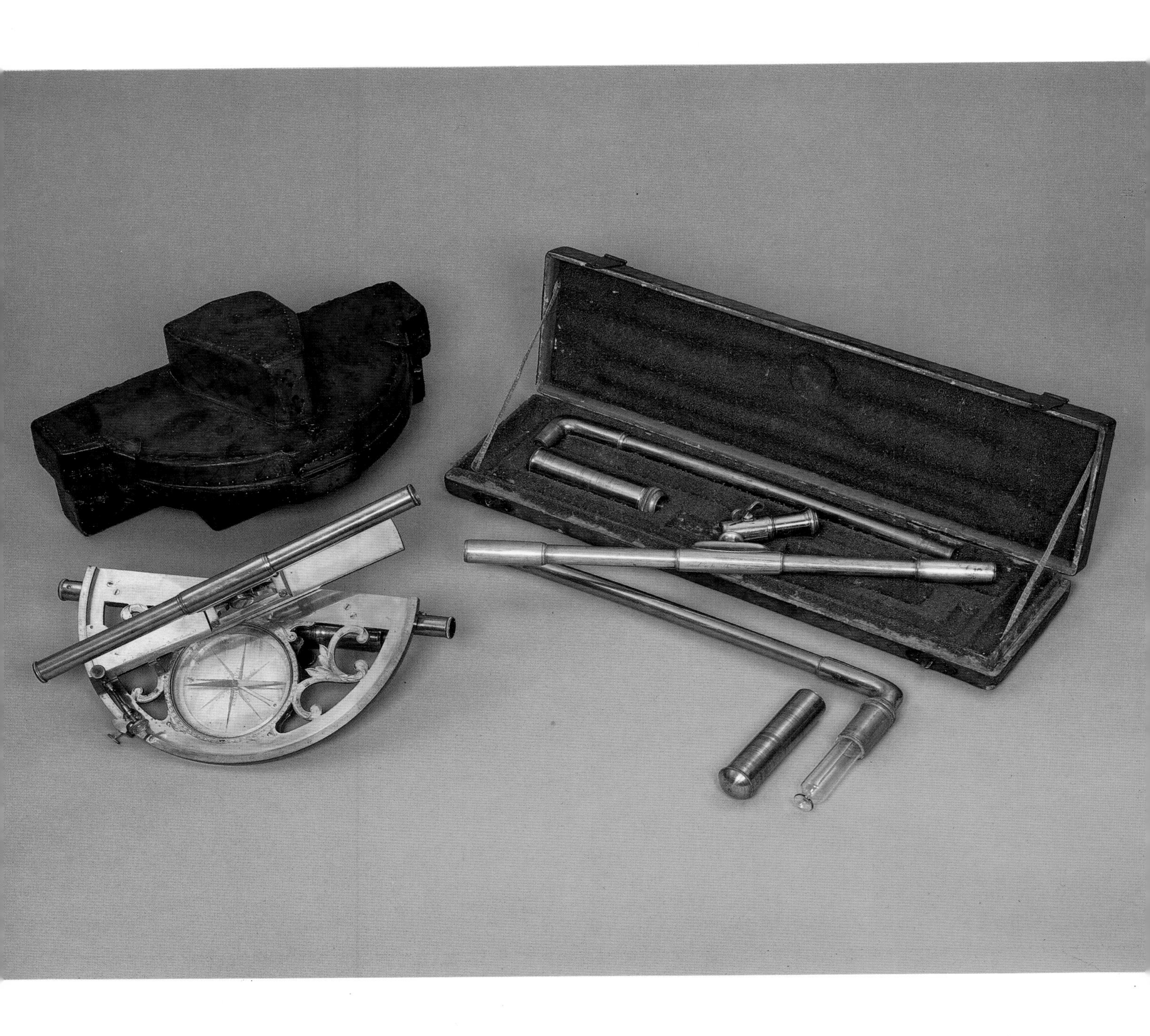

98 Graphometer and water level by Lennel, both 1774, radius of the former 155 mm, length of the latter 1.36m.
The graphometer is signed 'Lennel Elêve et Successeur de Mr Canivet à la Sphére AParis'. The water level depends on the fact that two connected columns of a liquid will find the same level.
Edinburgh, Royal Museum of Scotland.

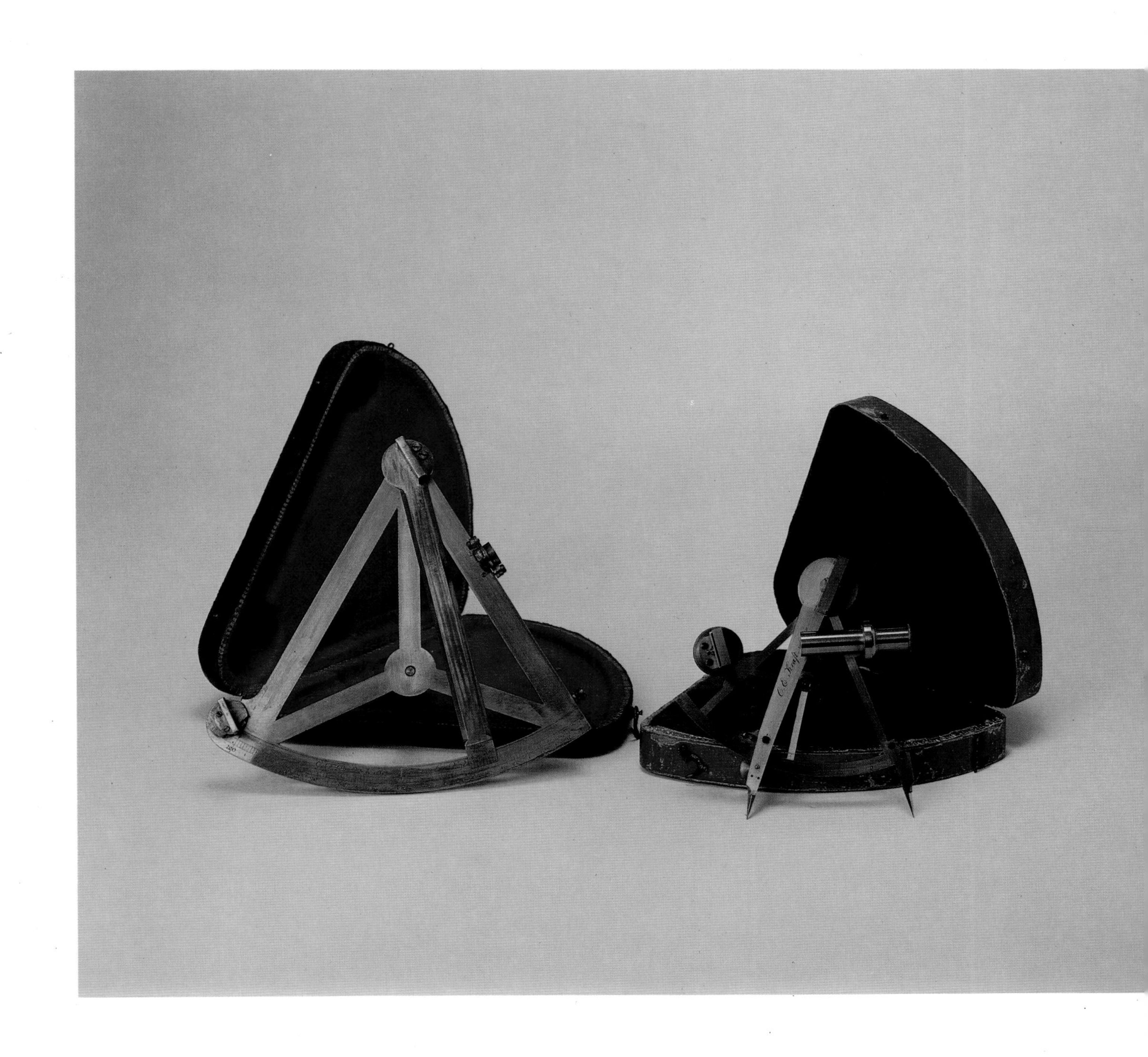

99 Two late 18th-century surveying sextants: left, by Brander & Höschel, radius 159mm; right, C. E. Kraft of Vienna, radius 87mm. The index arm of the Kraft sextant also forms one leg of a pair of dividers. Cambridge, Whipple Museum.

100 (*above*) Circumferentor or miner's dial by Cary, *c*.1800, length 290mm. A standard circumferentor with a transverse bubble level, but with the optional addition of a vertical semicircle with a longitudinal bubble, attached to the circumferentor by knurled screws. Cambridge, Whipple Museum.

101 (*right*) Graphometer by Paul of Geneva, 18th century, with the usual compass, plain sights and ball-and-socket joint, but with a removeable vertical arc, telescopic sight and bubble level attached by knurled screws to the moving alidade. Geneva, Musée d'Histoire des Sciences.

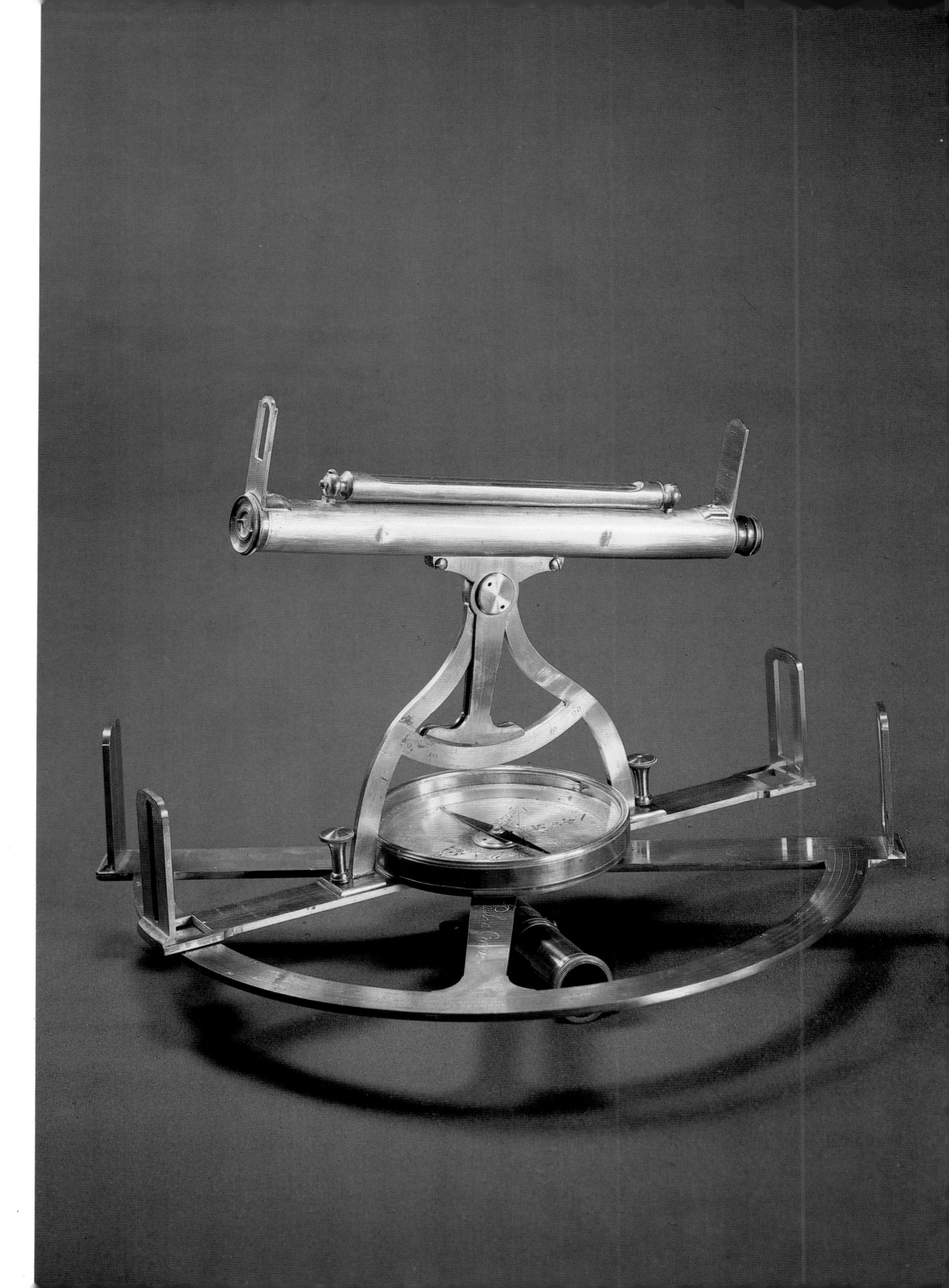

102 (*below*) Azimuth compass by Nairne, *c.*1765, diameter (card) 165mm, with pivoted sights and a gimbal mount. Harvard University, Collection of Historical Scientific Instruments.

103 (*right*) Theodolite by Heath & Wing, *c.*1740, diameter (horizontal circle) 304mm. This is an example of a combined simple and altazimuth theodolite; the vertical arc can be removed, and replaced by fixed and pivoted plain sights. The horizontal circle (or simple theodolite) is signed 'T. Heath Fecit', and the vertical 'Made by Heath & Wing', suggesting that the instrument was upgraded after the partnership was formed in 1740. Cambridge, Whipple Museum.

104 (*above*) Brass sextant by Nairne, late 18th century, radius 160mm, divided to 30 minutes of arc and read by vernier to 1 minute; and a mahogany frame octant by Goater of London, *c*.1750, with a transversal scale on an inlaid boxwood arc, divided to 20 minutes and by transversals to 2 minutes. Cambridge, Whipple Museum.

105 (*right*) Portable equatorial by Nairne & Blunt, *c*.1780, telescope length 545mm. Nairne published an equatorial design in 1771 in *Philosophical transactions*, but a number of developments have been made. The partnership between Edward Nairne and Thomas Blunt lasted from 1774 to 1793. Cambridge, Whipple Museum.

106 (*above*) Reflecting circle by Troughton, *c.*1805, diameter 312mm, mounted on a stand, having been used in the Observatory of Cambridge University.
The silver scale is divided to 20 minutes of arc, and read by verniers on the three index arms to 20 seconds.
Cambridge, Whipple Museum.

107 (*right*) Transit theodolite or altazimuth instrument, by Thomas Jones & So c.1840, diameter (horizontal circle) 250mm, which might be used as a sophisticated theodolite or as an astronomical instrument. Both scales are divided on silver to 10 minutes of arc and read by vernier to 10 seconds; the cross-wires may be illuminated at night by light introduced through the horizontal axis; and the horizontal circle has an independent motion, with tangent screw adjustment, for repeating azimuth measurements.
Cambridge, Whipple Museum.

BRIDGES – LEE'S
PATENT
PHOTO THEODOLITE
J. H. STEWARD, 406 Strand, LONDON.
Nº 24.

7

The Growth of Observatories

GREAT MANY OBSERVATORIES WERE established in the eighteenth century; this is clear already from our outline survey, in Chapter 6, of the locations of English instruments. We referred to observatories in Bologna, Pisa, Milan, Padua, Florence, Naples, Palermo, Berlin, Göttingen, Mannheim, Gotha, Leipzig, Amsterdam, Gdansk, Eger, Madrid, Cadiz, Stockholm, Uppsala, St Petersburg and Vilna. All were eighteenth-century foundations. We can add British observatories in Blenheim, Cambridge (college observatories, the University Observatory was not founded till 1823), Kew, Oxford, Shirburn, Aberdeen, Glasgow, Dublin and Armagh. Most of the private foundations have been omitted, but those of the Duke of Marlborough (Blenheim) and the Duke of Macclesfield (Shirburn) were particularly well equipped. Other observatories founded in Europe were in Prague, Tirnan, Paris (the observatories of Delisle, Lalande, Le Monnier, Lacaille, Delambre, the Ecole Militaire and the Collège Royal), Giessen, Ingolstadt, Munich, Ochsenhausen, Wurzburg, Budapest, Lisbon, Alba Iulia (Romania) and Lund. We have noted only observatories with sizeable fixed instruments; many others were equipped with smaller, 'portable' instruments, but all needed to be supplied by specialist workshops. The growth in demand for precision astronomical instruments, by comparison with the needs of the handful of observatories active at the end of the seventeenth century, is astonishing.

The reasons for the blossoming of 'official' observatories – sponsored by states, provinces, churches, universities and scientific societies – were related, on the one hand, to practical needs, and on the other, to the contemporary image of science. The success of the Newtonian account of the planetary system, ever more clearly demonstrated by eighteenth-century mathematicians, established the picture of a world precisely managed on mathematical principles. It was understood that the empirical foundation for this picture derived from precision astronomy, and the consequent growth of observatories presented enormous opportunities to talented mechanics.

108 (*opposite*) Photo theodolite by J. H. Steward, early 20th century, diameter (horizontal circle) 198mm, with an aluminium camera mounted on a divided horizontal circle and surmounted with a telescopic sight and vertical arc. A compass within the camera box may be read from behind, and its image is reproduced on the photographic plate. Cambridge, Whipple Museum.

The mural quadrant

The most ubiquitous of the large observatory instruments of the eighteenth century was the mural quadrant, and the most influential model (fig. 110) was that made by George Graham for Edmond Halley at Greenwich. It was mounted in 1725, and in 1738 was considered worthy of an entire chapter in Robert Smith's standard textbook, *A compleat system of opticks* (Cambridge, 1738).

Its immediate predecessors were Flamsteed's mural instruments, particularly the 10 ft quadrant designed by Hooke and built by Tompion, later Graham's partner in clockmaking. The most obvious improvement was the much stronger bracing, to give rigidity to the frame and so hold the limb and telescope mount in their proper orientation. This frame was composed of iron bars, most of them set in the plane of the quadrant, but reinforced by others perpendicular to the plane, and all riveted together by right-angle brackets. The bars ran in four directions – vertical, horizontal and along both diagonals of the squares so formed. This framework was constructed by Jonathan Sisson, under Graham's direction.

The limb, of 8 ft radius and 3½ in wide, was of brass and divided by Graham himself. Here again there was notable innovation. The limb was divided twice, the inner scale being a conventional one of 90°, subdivided to 5 minutes. The outer scale was divided into 96 parts – a more convenient and accurate division, since, unlike 90 parts, it could be achieved merely by successive bisection. First the 60 degree (64–part) arc was established, by striking off a chord equal to the radius of the quadrant. This was bisected successively to 32, 16, 8, 4, 2 and 1 parts, and then subdivided to 16 'partlets'. The remaining 30 degree (32–part) arc was begun by striking off a chord established by the first bisection of the larger arc. Tables could then be used to check the conventional division of 90° against that of 96 parts.

The quadrant was mounted on a pier built of nine stones. The telescope was counterpoised for ease of motion, and had a fine ad-

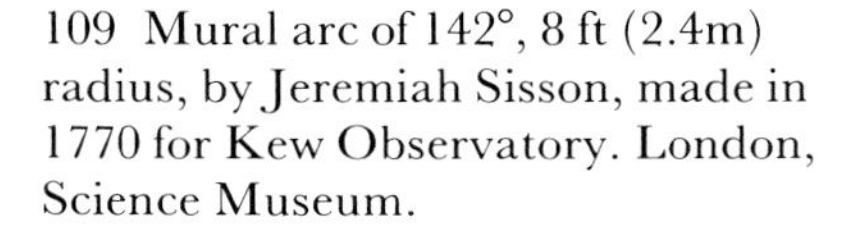

109 Mural arc of 142°, 8 ft (2.4m) radius, by Jeremiah Sisson, made in 1770 for Kew Observatory. London, Science Museum.

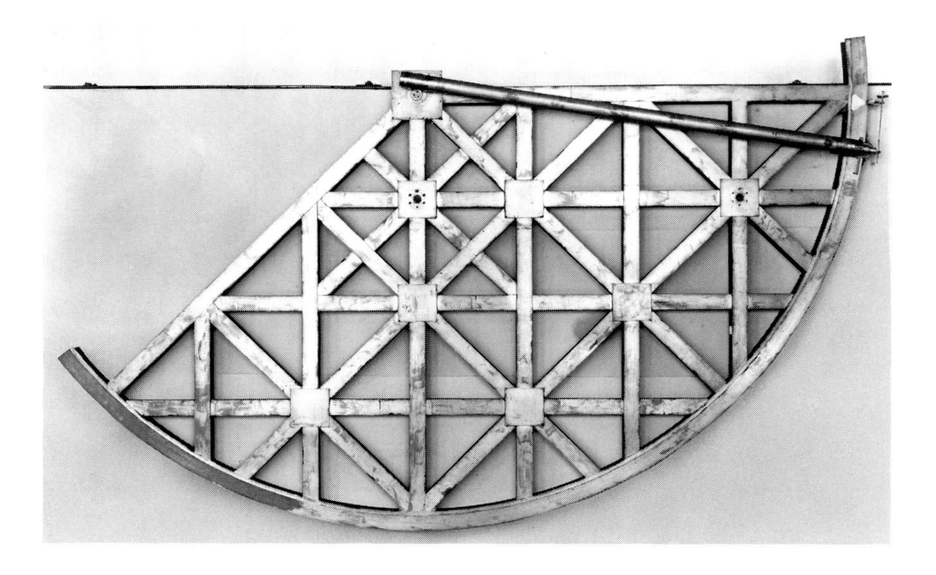

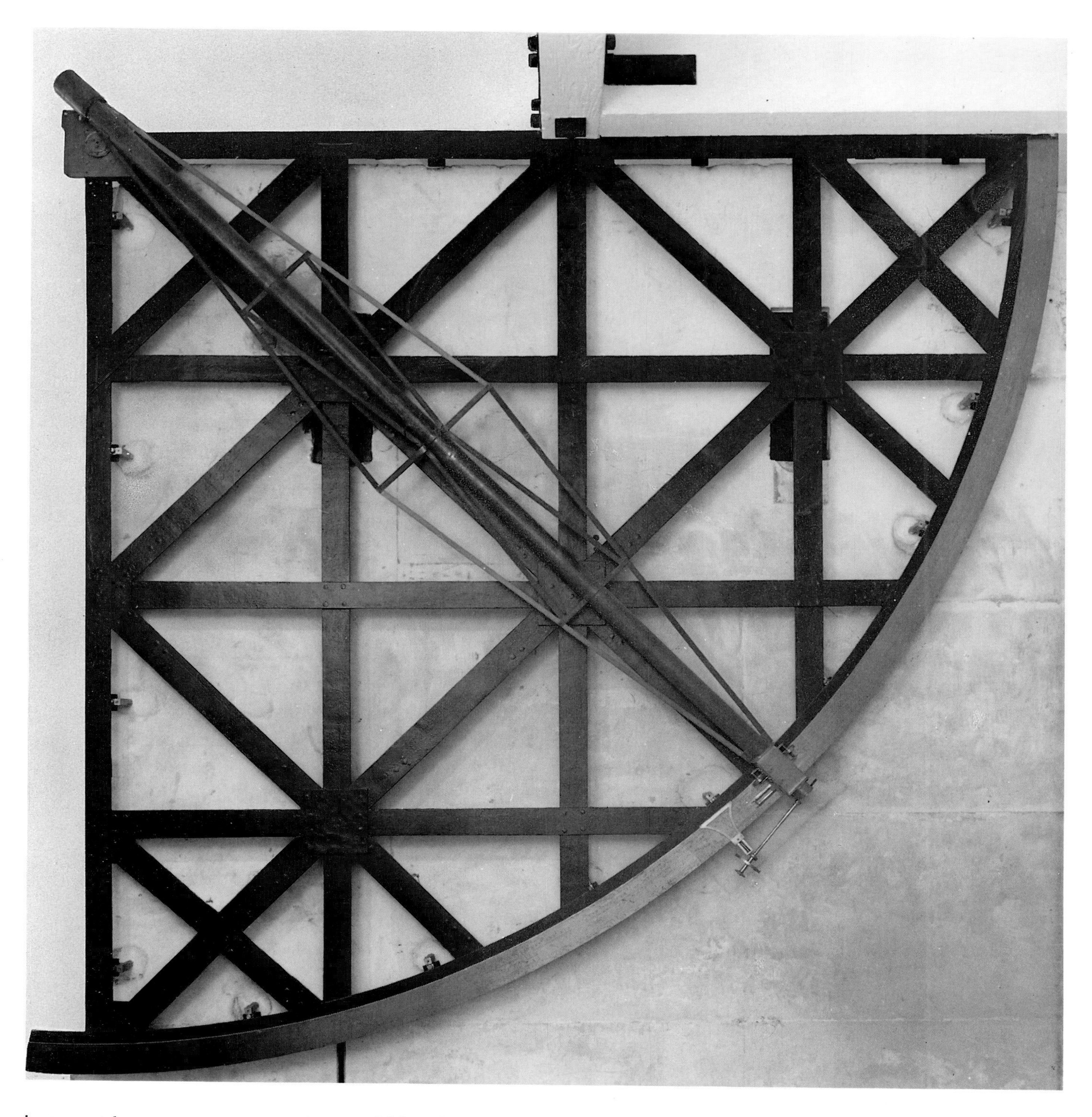

justment by a tangent screw that could be clamped to the limb. By means of a vernier, the final reading was accurate (in theory) to half a minute. Subdivision by the vernier of the 96–part scale was to 1/16 of the scale division, i.e. to 1/256 parts.

A fixed quadrant was, of course, restricted to half the sky (the southern half in this case), and Smith pointed out that funds should be found for completing the work, now that its usefulness had been established 'by an incredible number of observations made with this noble instrument by our Royal Astronomer Dr. Halley . . . [close to

110 The 8 ft (2.4m) mural quadrant at Greenwich by Graham, completed for Halley in 1725 (cf. figs. 109, 111, 112). The frame was constructed by Jonathan Sisson under Graham's direction. The tangent screw for fine adjustment is visible, but the original counterpoise is missing. Greenwich, National Maritime Museum.

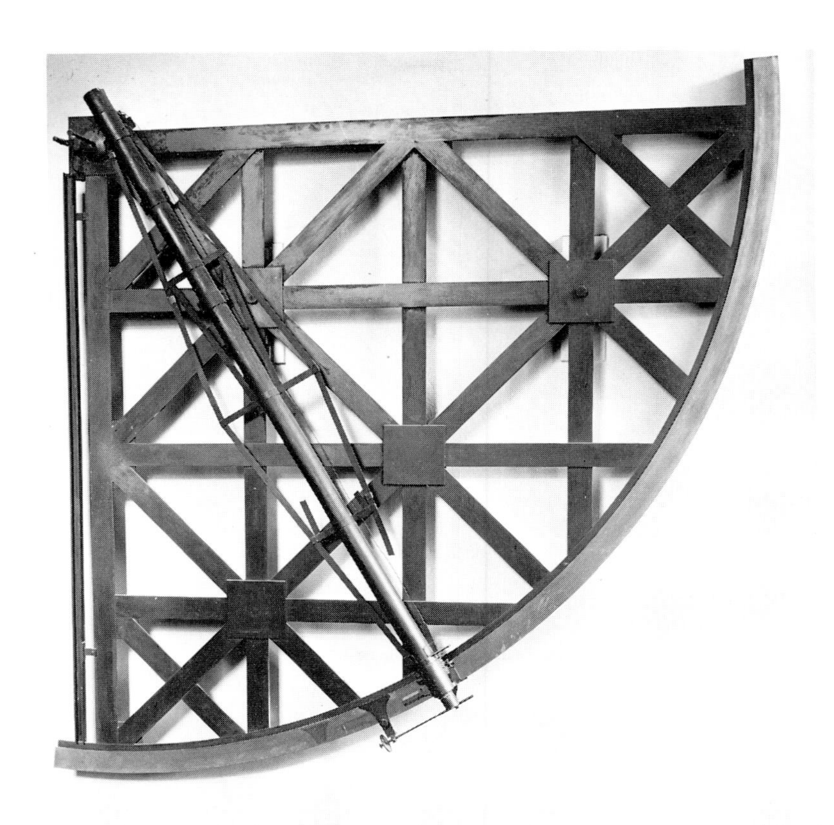

111 Mural quadrant, 1772, radius 8 ft (2.4m), by Bird, one of a pair made for the Radcliffe Observatory. The tube on the left is for the plumb-line. Oxford, Museum of the History of Science.

112 (*below*) Mural quadrant by Jonathan Sisson, 1739, radius 1.65m, with the original form of counterpoise. Bologna, University Observatory.

a] . . . sufficient exactness for finding the Longitude at Sea' (Smith, 1738, p.332):

> When this quadrant was fixt upon the eastern side of the wall, some provision was made for fixing such another upon the western side; to observe the altitudes and transits of stars over the northern half of the meridian; and as soon as the Government shall think proper to allow the expence of so useful an instrument, we may expect in a few years time to be furnished with a more exact Catalogue of all the stars visible in our hemisphere, than can possibly be had from all the instruments and observations yet extant. And such a Catalogue being the very foundation of all the desired accuracy in Astronomy, Geography and Navigation, I need not dilate upon the usefulness of it. (*ibid*, p.341)

In 1750, under the new Astronomer Royal, James Bradley (1693–1762), Bird indeed built a second quadrant – a copy of Graham's, except that the frame was in brass. A brass frame added to the expense of the instrument, but Graham's quadrant had gradually distorted under its own weight. Bird described his quadrant and the method of division in two pamphlets published by the Board of Longitude in 1767–8, where he explained how he had begun in 1748 by studying the model left by Graham:

> . . . I made myself fully acquainted with the general construction of the old Quadrant, which was executed under the direction of the late Mr. Graham, and found the general plan, though little taken notice of at that time, to be such, as, I think, will be a lasting testimony of his great skill in mechanicks. (Bird, 1768, p.7)

A dissenting opinion from an influential source was recorded in 1758, in Edmund Stone's Supplement to the second edition of his translation of Bion's *Construction and principal uses of mathematical instruments*. While acknowledging that Graham had been 'one of the greatest Masters of Mechanicks in the World', he considered that his quadrant was 'too complex, and redundant in Contrivance, both as to Strength and Exactness' (Bion, 1758, p.278). In other words, he considered that Graham had been aiming too high – for an inappropriate and spurious level of accuracy. The subsequent history of instrumentation has proved Stone wrong, and the working astronomers of the second half of the century were convinced by Graham's work, as publicized by Bird. We have seen that a great many such quadrants were made for European observatories, mostly by either Sisson (fig. 112) or Bird (fig. 111).

Smith's enthusiasm for twin mural quadrants, as the instrumentation sufficient to re-establish the observational basis of precision astronomy, derived from the fact that both he and Halley believed the quadrant could fix stellar positions in two co-ordinates. Altitudes (and thence declinations) were measured directly, while right ascensions could be found by timing transits. In fact, for a variety of reasons, the two functions came to be separated in the second half of the century, and a specialized instrument more commonly used for right ascension.

The transit instrument

The transit instrument (fig. 113) was not, of course, a new invention. Römer had made use of both a simple transit instrument, used only for measuring right ascension by a clock, and a full transit circle, which could take altitudes (and declinations) as well. In the eighteenth century, the transit instrument came into its own, but after a considerable period of uncertainty. Only in the late century did a transit instrument come to define the primary meridian of most observatories.

In fact, Greenwich was early in being so equipped, since Halley had a transit instrument in 1721, four years before his quadrant. He may have regarded it as a temporary expedient, for he abandoned it as soon as the quadrant was in use. Lalande linked this early instrument with Hooke, although Hooke had died in 1702.

The telescope was 5½ ft long, and was mounted towards one end of the 3½ ft axis, after the manner of Römer. The supports have not survived, but were probably mounted on two stone piers, and carried the cylindrical end-pieces of the axis. In 1738 Smith described bearings for such an instrument as V-shaped slots cut in thick brass plates, and these bearing plates as movable – one in the horizontal, the other in the vertical direction – with respect to two supporting fixed plates. The plate with a horizontal motion was used to set the axis perpendicular to the meridian, and so to correct for what is known as 'azimuth error'; the one with a vertical motion was for setting the axis level, and Smith describes a hanging spirit level for this purpose. A third error, the 'collimation error', occurred when the

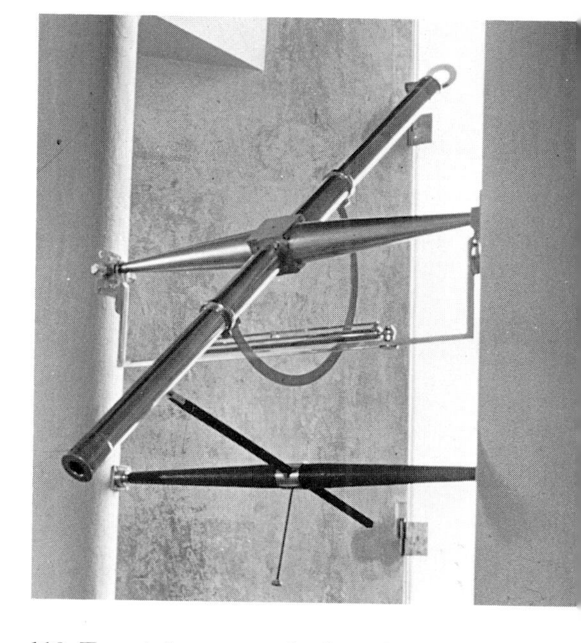

113 Transit instrument by Jonathan Sisson, 1739, telescope length 1.38m, with a hanging level and illustrating the typical double-cone form of axis. This is not the original mounting. Bologna, University Observatory.

optical axis (the line through the centre of the object glass and the intersection of the cross-wires) was not perpendicular to the axis, and was corrected by adjusting the wires.

By Bradley's time, it was clear that adjustments necessary to maintain accuracy in the altitude function of the mural quadrant were not readily combined with those required for right ascension. He decided to separate the two functions in two instruments, and in 1749 ordered a new transit instrument from Bird. It had an 8 ft tube and a 4 ft axis, and was considerably closer to the design of instrument that became standard for the remainder of the century. Whereas Halley's telescope had been braced by struts connected to the axis, the axis itself was now composed of two cones, with the telescope carried by their bases and placed centrally between the two supports.

The cross-wires of the transit instrument required illumination, particularly at night. This was achieved simply by introducing light from a candle or oil lamp directly through an aperture in the tube. During the daytime, a reflector, generally in the form of an elliptical ring, could be positioned beyond the end of the tube without restricting the useful portion of the object glass.

The astronomical sector

The term 'sector', as applied to astronomical instruments, has been used to describe those whose arcs do not fit the standard categories of octant, sextant and so on, and it is applied in that sense to certain instruments of Tycho, Hevelius and Flamsteed. In the eighteenth century, it came to refer to two types of instrument characteristic of the period, both of which were originally designed by Graham: the zenith sector (fig. 114) and the equatorial sector. The former was much the more common, being a fairly regular item of observatory equipment, perhaps on account of the celebrity of the results obtained with Graham's original.

Two zenith sectors are recorded at the Paris Observatory early in the century, but the standard design derives from an instrument built by Graham for Bradley's private observatory in 1727. Bradley brought it with him to Greenwich in 1749. We have seen that Hooke had a zenith instrument mounted in his rooms in Gresham College, in his attempt to detect an annual stellar parallax and so prove the motion of the Earth. This instrument was 36 ft long and was fixed, with measurements being taken by an eyepiece micrometer. Samuel Molyneux (1689–1728), a friend and collaborator of Bradley, had a private observatory at Kew, close to London, and was interested in pursuing Hooke's methods. He obtained a 24 ft zenith instrument from Graham in 1725, and it was with this instrument that Bradley began his own search for stellar parallax.

Bradley's improved sector was approximately half the length of Molyneux's, but had a range of some 12½° across the zenith. It was suspended about a horizontal axis close to the upper end of the tube, with the arc attached to the lower end and read by a plumb-line. The motion of the tube in the meridian was controlled by a micrometer

114 Zenith sector by Bird, 1772, radius 3.6m, made for the Radcliffe Observatory and modeled on Graham's instrument at Greenwich. The tube is suspended at its upper end on a horizontal axis, the arc moves with the tube and is read by a plumb-line. Oxford, Museum of the History of Science.

screw, whose graduated head subdivided the 5–minute divisions of the arc to one second.

While Bradley's programme of observations with this instrument was aimed at detecting stellar parallax, he found an annual deviation in stellar positions in the sense opposite to that expected. He eventually explained these results in terms of what is called 'stellar aberration' – an effect of combining the velocities of light and of the Earth's orbital motion. Though the result was not anticipated, it was no less a proof of the motion of the Earth, and was accepted as the first such demonstration. Bradley went on to detect 'nutation' – a periodic revolution of the Earth's axis – with the same instrument.

The equatorial sector, or 'Mr Graham's astronomical sector' as Smith described it in 1738, allowed any small separation to be measured in right ascension and declination. A frame incorporating an arc of some 10° and a pivot for the telescope motion were attached to a circular plate. This plate was held parallel to a polar axis and could be clamped for any range of declination. By rotating and then clamping the polar axis, the arc could be brought into any right ascension. The sector was suited to small angular separations that were nonetheless too large for the micrometer, differences in right ascension being taken from a clock and in declination from the arc. Once again the design was by Graham, and Bradley's first sector became a model for later instruments.

When the Radcliffe Observatory was established at Oxford University in 1772, the major instruments ordered by Thomas Hornsby (1733–1810) from John Bird were two mural quadrants after Graham's design (fig. 111), a transit instrument and equatorial and zenith sectors (fig. 114). It is clear that Graham had been the dominant figure in the reform of observatory instruments, and Bradley generously acknowledged his debt to an outstanding mechanic:

> I am sensible, that if my own Endeavours have, in any respect, been effectual to the Advancement of Astronomy, it has principally been owing to the Advice and Assistance given me by our worthy Member [of the Royal Society] Mr. George Graham; whose great Skill and Judgment in Mechanicks, join'd with a complete and practical Knowledge of the Uses of Astronomical Instruments, enable him to contrive and execute them in the most perfect manner. (*Philosophical transactions*, 45, 1748, p.6)

When the Radcliffe trustees next ordered a major instrument, however, it was a design unknown to Graham – a meridian circle, built by Thomas Jones in 1836. Observatory instruments had again undergone an important reform, begun by a new generation of mechanics in the late eighteenth century.

Portable instruments

Before considering the genesis of the large observatory circle with telescopic sights, whose story extends into the nineteenth century, we should note the growing family of portable instruments, derived from their larger observatory counterparts. The typical portable instrument of the eighteenth century was the quadrant – sufficiently well established in 1709 to be the only astronomical instrument described and illustrated in Bion's *Traité de la construction et des principaux usages des instrumens de mathématique*. This example had a single fixed sight on the upper radius; the entire quadrant thus moved in altitude with the telescope, and the reading was taken from a plumb-line (cf. fig. 116).

The same illustration appears in the first English edition of Bion's textbook in 1723. In the 1758 edition, however, where Stone appends descriptions of English instruments, he illustrates a 3 ft quadrant where readings are taken, not by a plumb-line, but by a telescopic sight moving over the limb (cf. fig. 65). This was a characteristic difference between French and English quadrants. Stone's instrument in fact had two sights: one fixed to the upper radius, the other linked to an endless screw, which gave a fine adjustment when racked along the toothed edge of the limb. The frame itself had two adjustments by rack and pinion, so as to move the quadrant into any plane for

115 (*right*) Portable astronomical quadrant by Jonathan Sisson, 1739, with an unusual frame based on a circle with a mahogany T-shaped support. It seems that the readings were taken by a plumb-line, which helps to explain the unusual frame, since, if so, the whole quadrant had to rotate in altitude (cf. fig. 116). Bologna, University Observatory.

116 (*opposite*) Astronomical quadrant by Butterfield, early 18th century, radius 640mm. A quadrant of the French type, with a plumb-line and single fixed telescopic sight. London, Science Museum.

A PARIS

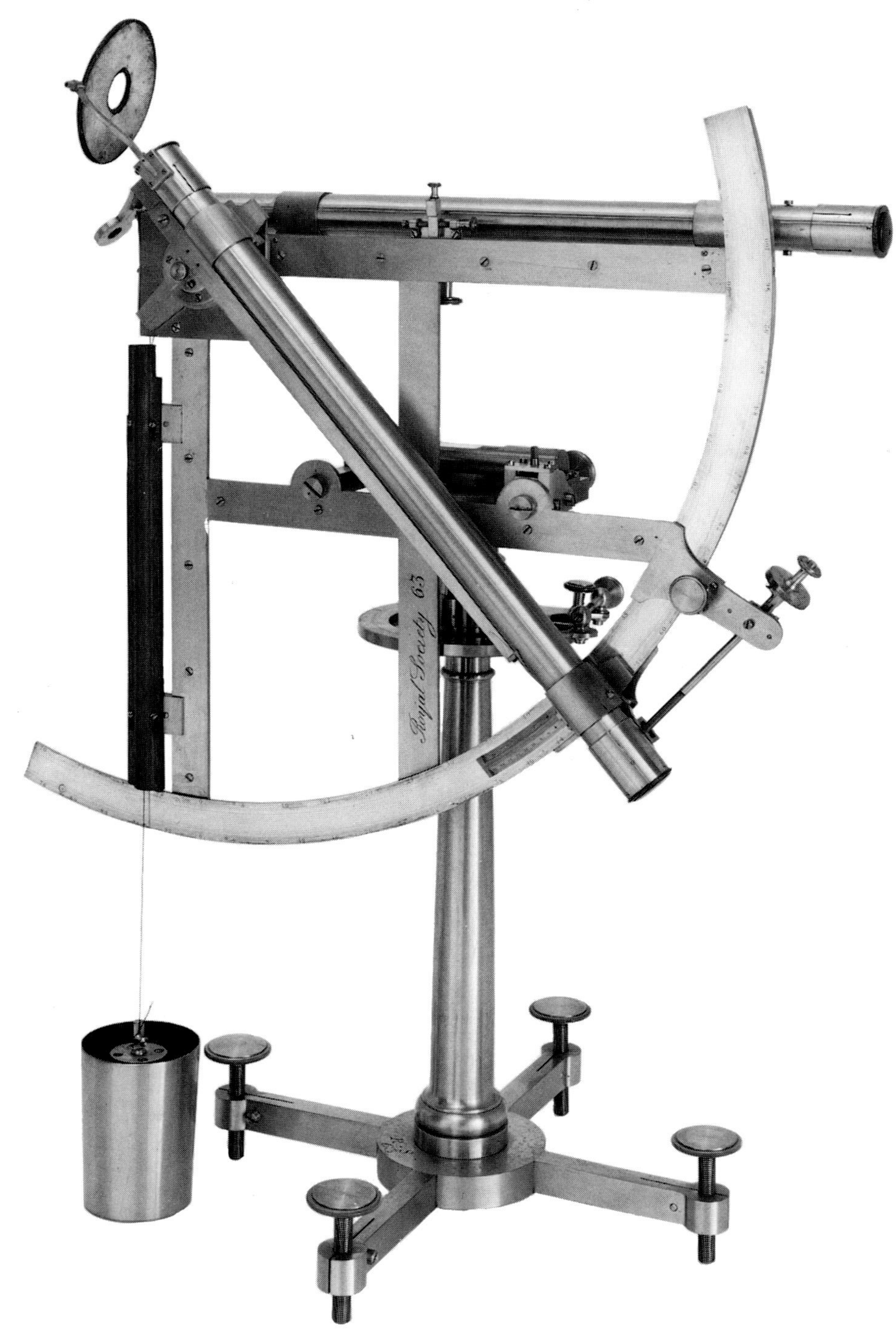

117 Portable astronomical quadrant by Bird, *c.*1767, radius 12 ins (305 mm), made for a 1769 Transit of Venus expedition mounted by the Royal Society. With two telescope sights (fixed and pivoted) on this quadrant, the plumb-bob is for levelling; it is suspended in water to damp its motion. There are two divisions of the quadrant, to 90° and 96 parts, with a double vernier plate for reading both scales. The elliptical ring beyond the object end is for reflecting light to illuminate the cross-wires. London, Science Museum.

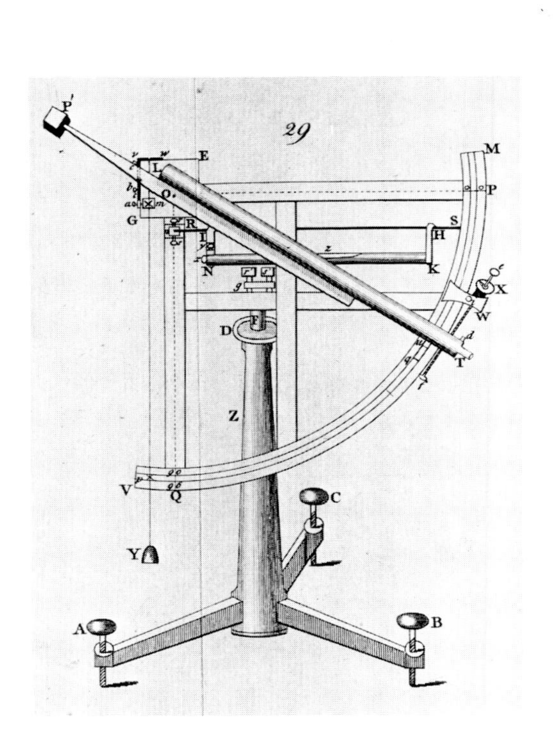

118 Ramsden's portable quadrant illustrated in Vince, 1790.

taking angular separations. Portable quadrants with a number of these features could be had from John Bird.

English and French designs were both current throughout the century. In 1790 Samuel Vince (1749–1821) illustrated the portable quadrant by reference to an instrument of Ramsden's (fig. 118). This had a single sight moving over the limb, and fine adjustment and subdivision by clamp and tangent screw. Levelled by spirit-level or plumb-line, it could move in azimuth and be inverted over a horizontal axis. Its motions were thus more restricted than the Bird/

Stone example, but were adequate for the kind of uses that might reasonably be made of a portable quadrant, such as latitude and time determinations.

On the French side, an instrument (fig. 119) constructed on the same general principles as that of Bion is illustrated in Lalande's *Astronomie* of 1764, and the same plate is included in the revised third edition of 1792. Here the text has been altered, but the portable quadrant is still described as:

> . . . de tous les instrumens actuels d'astronomie que nous avons à décrire, celui dont l'usage est le plus général, le plus indispensable, le plus commode. (Lalande, 1792, vol. 2, p.580)

In fact, in the later part of the century, the French introduced the reflecting circle as a portable astronomical instrument, and it quickly gained in popularity. We can leave it aside for the present, however, as it was equally well adapted to astronomy, navigation and surveying, and most surviving instruments are of the hand-held navigation type.

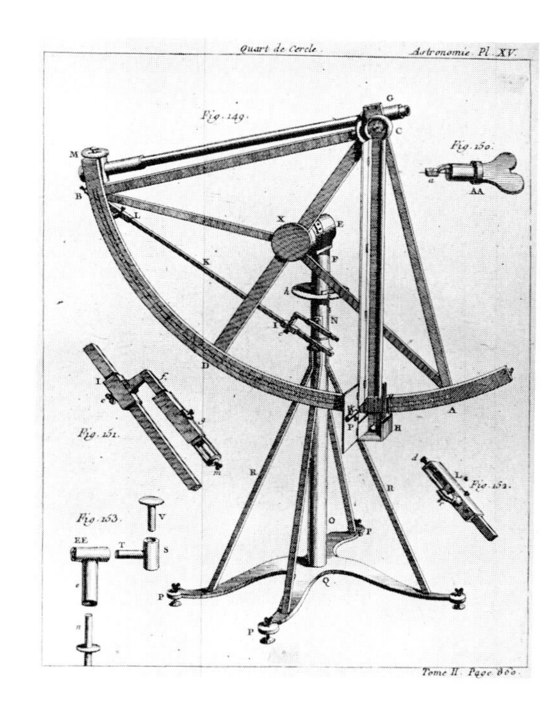

119 The portable quadrant design included in various editions of Lalande's *Astronomie.* (Lalande, 1764)

If the portable quadrant represents merely the final adaptations of some very old designs, elsewhere the smaller instrument served as a testing-ground for new ideas. Most of the equatorial instruments manufactured in the eighteenth century were portable. We have seen that large equatorials had been built, but they had not been capable of the accuracy and consistency of the more basic altitude and azimuth instruments. The problem had been to find a steady and reliable mounting, and experiments with different types of mount were for a time mostly confined to the small scale.

These portable equatorials also met a new consumer demand. The eighteenth century was marked by a growth in popular interest in science and the foundation of many private observatories. Few had the means to equip their observatories in the style of those at Blenheim or Shirburn Castle, but the portable equatorial could supplement the ordinary refractors or reflectors, used simply for observing. Serious amateurs could make positional measurements of notable objects, or could use the divided circles as setting circles, to find objects they had heard of or read about. The equatorial was much better suited to such amateur work than the more fundamental quadrants or transit instruments.

The situation is nicely epitomized by an early portable equatorial by Short, whose design (fig. 120) was published in the *Philosophical transactions* for 1749. A base with four levelling screws carried an azimuth circle on four pillars, and above this four pillars carried the horizontal axis of an altitude or latitude arc (its main function was to set the instrument for latitude) of some 270°. This arc carried a further plate, which could thus be set parallel to the equator and was divided into hours of right ascension. A further four pillars carried the axis for a declination arc, surmounted by the telescope, and each of the four motions – azimuth, altitude, right ascension and declination – was given a fine adjustment by rack and pinion. After this demonstration of positional expertise, the telescope itself was a Gregorian reflector (Short made only reflectors), which was really suited only to observing and not to positional work.

120 Short's equatorial mounting for a Gregorian reflector. (Short, 1749)

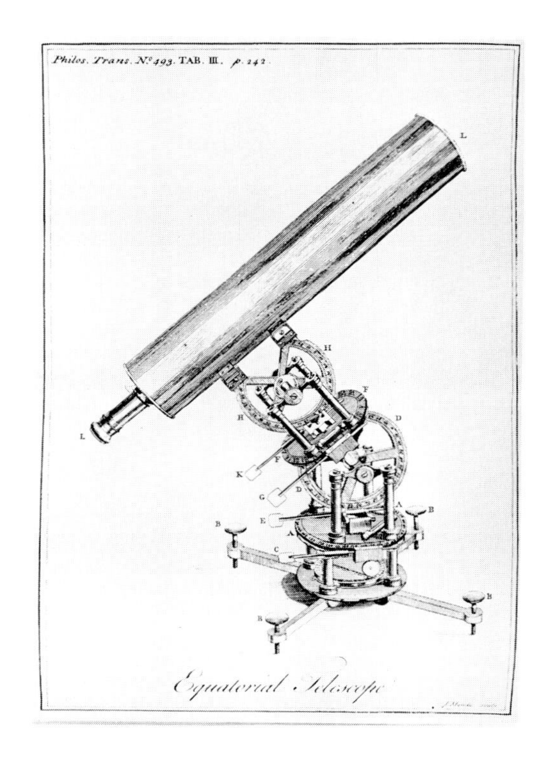

121 (*left*) Portable equatorial by Dollond, *c.*1800, of a design similar to Ramsden's (fig. 123), but without the refinement of conical bearings. London, Science Museum.

122 (*right*) Portable equatorial instrument by Adams, *c.*1785, telescope length 291mm. The telescope, in Y bearings, moves over an 'inverted' arc, mounted on an equatorial circle, itself carried by a latitude arc, whose axis rests on two A frames supported by the horizontal circle. The four circles or arcs are divided, respectively, for declination, right ascension, altitude and azimuth. Cambridge, Whipple Museum.

This general arrangement of divided circles, reminiscent of the medieval torquetum, was typical of a series of portable equatorials designed in the second half of the century. Azimuth and altitude arcs or circles were used to set up the instrument, right ascension and declination to find objects or record positions. However, they appeared in different guises, according to the variations attempted by the makers. In England, designs surmounted by refractors were devised by Nairne (fig. 105), Ramsden (fig. 123) and Dollond (fig. 121) in the 1770s. Towards the end of the century Troughton built an instrument of this type (fig. 124) on such a large scale that, although universal in the sense that it could be adjusted for latitude, it was scarcely portable. It stood 7 ft high and was an attempt to carry this progression of designs on to a genuine observatory instrument.

The more complex versions could be frustrating to adjust. Vince chose to describe Ramsden's design in his textbook of 1790, and in 1828 Pearson commented on the effects of Vince's account in his usual dry fashion:

123 (*left*) Portable equatorial by Ramsden, late 18th century, with conical bearings for the horizontal and equatorial plates, the latter bearing being adjustable at its lower end along a latitude arc. The axis for the declination arc is carried by two A supports on the equatorial plate. London, Science Museum.

124 (*below*) Troughton's Coimbra equatorial: an attempt to build a 'portable' equatorial appropriate to a major observatory. (Pearson, 1828)

> When Professor Vince received, from Mr. Troughton, instruction how to proceed with the adjustments of Ramsden's equatorial, he followed the directions only in part, and so far puzzled himself with his own methods, that he has not failed to puzzle also those amateurs in practical astronomy, who have been taught by him to undo one thing by doing another. (Pearson, 1828, p.521)

In fact, the most successful design in this class of instrument was the 'portable observatory' (fig. 122) described in the popular works of George Adams, Jr. An azimuth circle with a magnetic compass – used to set the instrument in the meridian – carried the axis of a latitude semicircle, surmounted by a right ascension circle and, in turn, a declination semicircle. Only the right ascension circle might be supplied with rack and pinion, and in general the flimsy mounting had no pretensions to carrying a serious astronomical instrument. Rather, it met the requirements of astronomy as a hobby.

The attitude of the professional towards attempts to elevate the equatorial into a positional instrument is again epitomized by Pearson:

> . . . notwithstanding the convenience with which a star or other body may be found by the equatorial motion, when the right ascension and

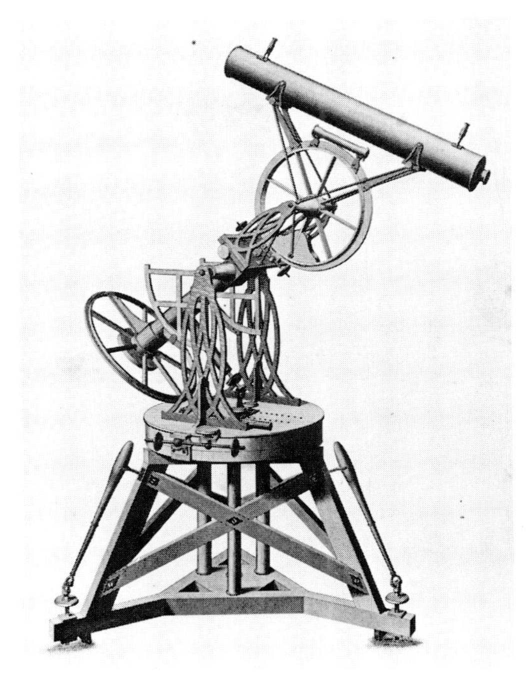

125 Equatorial, 1791, polar axis 2.5m, made by Ramsden for Sir George Shuckburgh. This type of mount is known as the 'English' equatorial. London, Science Museum.

> declination of the body are previously known, astronomers are not generally disposed to place much confidence in the accuracy of observations, for determining unknown right ascensions and declinations, out of the meridian, however well the instrument may be equipoised. (Pearson, 1828, pp. 523–4)

This is not to say that no large equatorials were built, and one in particular, even though it was not a success, is worth noting.

Built by Ramsden for the amateur astronomer Sir George Shuckburgh, it was finished in 1791, ten years after it had been ordered (fig. 125). A polar axis over 8 ft long was carried by two detached supports and had a declination circle of 4 ft diameter at its lower (south) end. From there a cone rose to a circular plate carrying six pillars, which enclosed the 4 ft diameter right ascension circle, and which carried its axis. This circle was double; the two elements were separated by pillars and enclosed the 4.1 in aperture telescope.

Though transferred to the Royal Observatory in 1811, the instrument was never a success, but it was significant in two respects. Firstly, it was the earliest large equatorial telescope. Secondly, a modified arrangement of pillars enclosing a double circle was used by Ramsden in the very important altazimuth circle he had completed for the Palermo Observatory by the time of its foundation in 1790.

The divided-lens micrometer

A great many types and variations of micrometer were designed in the late eighteenth and early nineteenth centuries. A full account would not be appropriate here, and they are described in detail by Pearson. But we cannot ignore a device so original as the divided object-glass micrometer (fig. 126), invented by John Dollond and described in the *Philosophical transactions* for 1753.

If the object glass of a telescope is split into any number of portions, which are then displaced from each other while remaining in the same lateral plane, each one will form a complete image correspondingly displaced in the focal plane. While each image will be complete, its intensity will depend on the size of the fragment.

In practice, the lens is carefully divided along a diameter and each half set in a separate brass mount. These may be displaced along the diameter by a double rack and pinion, moved from the eye-end by means of a universal joint. The amount of displacement can be measured by a linear scale with a vernier, once, for example, two stars that are normally seen separate have been made to coincide, or the opposite edges of a planet's disc have been brought together. A second handle and universal joint rotates the object glass by rack and pinion, so that measurements can be made in any direction.

The divided object-glass micrometer was first applied, in fact, to the reflecting telescope, by means of a long focal length lens attached to the object end of a Gregorian or Cassegrain. It was sold as an accessory with the more serious telescopes of Short, though the lenses were by Dollond. It is in this form that such micrometers are most commonly found today, and instruments of this type are sometimes called 'heliometers'. They were well suited to solar measurement, because they displayed a complete image – or, rather, images – of the Sun's disc.

Later, it was found possible to adapt divided lenses to the objectives of refractors, and to construct eyepiece micrometers based on

126 Divided object-glass micrometer, late 18th century, aperture 128mm. The scale for measuring the displacement between the half-lenses is 5.5 ins long, divided to 0.05 ins and read by vernier to 0.002 ins. Cambridge, Whipple Museum.

the divided lens technique. The latter most commonly took the form of dynameters – used to measure a telescope's magnifying power – and the original design was due to Ramsden, but to follow this progression would carry us too far into the realm of optics.

The early circle

The most important development of the late eighteenth century is represented by the first of the new generation of astronomical circles – full divided circles with telescopic sights. Capable of more accurate division and more accurate reading, and susceptible of techniques for reducing errors, they quickly established, in the early nineteenth century, a new pattern for the fundamental instruments of astronomy. The first crucial designs, however, appear at the end of the eighteenth century in the work of the most able makers. Pearson described the situation in this period as follows:

> Mural quadrants had begun to sink in public estimation, and circles had been proposed to supersede their use . . . as being more capable of being accurately divided, and of having their index errors appreciated, as well as the errors of excentricity corrected by opposite readings.
>
> (Pearson, 1828, p.402)

127 (*above*) Ramsden's altazimuth built for the Palermo Observatory in 1790 There were four pillars set on a marble stand, each rising to support the top of the axis; for the sake of visibility the illustration shows only two. (Pearson, 1829)

A full circle with a telescopic sight was not entirely new. We have seen that Römer had a meridian circle in Copenhagen. Johann Jakob Marinoni (1676–1755) established a private observatory in Vienna, equipped from 1740 with two 9 ft mural quadrants, a 6 ft transit circle and other instruments made to his own designs. A number of small meridian circles and portable altazimuth circles were also made. But the celebrated Palermo instrument (fig. 127) by Ramsden represents the true beginning of a succession of large observatory circles.

Ramsden's altazimuth was one of the observatory's two foundation instruments in 1790; the other was a transit instrument, also by

128 (*right*) The 8 ft (2.4m) altitude circle, designed and begun by Ramsden and finished in 1809 by Berge, illustrated in the Observatory of Trinity College, Dublin, with the transit instrument. (Taylor, 1845)

Ramsden. The vertical axis of the altazimuth was supported by two metal arches, crossing at right angles above the instrument, and carried by four metal pillars set on a marble stand. At the lower support, the apex of an inverted cone carried the 3 ft diameter azimuth circle, divided to 6 minutes and read by a micrometer microscope to a single second. The microscope served to magnify the scale, and its eyepiece micrometer to measure the position of the index between two divisions. There was thus no need for a vernier scale.

The cone rose to a rectangular plate, carrying four 6.5 ft pillars that supported and enclosed the 5 ft altitude circle. This was a double circle, the two elements secured to the telescope and separated by a series of bars. Eight of these bars were attached to conical struts radiating from the 2 ft horizontal axis. The vertical circle also was divided to 6 minutes, and was read by two micrometer microscopes – at opposite ends of the vertical diameter – so as to incorporate a correction for eccentricity.

The circle was a success and was put to good use by Piazzi, who quickly published a full account of it. He also gave it a suitably impressive setting – a rotunda with pillars and an architrave of marble. Ramsden's best-known portrait shows him seated by his dividing engine – the invention that brought him the Royal Society's Copley Medal – and in the background is the Palermo circle.

An even greater circle (fig. 128) was designed and largely executed by Ramsden, but finished only after his death. This was an 8 ft diameter meridian circle, ordered in 1785 for the new observatory of Trinity College, Dublin. It was to be an altitude instrument only, to complement the 6 ft transit instrument also ordered from Ramsden and delivered in 1788, and was originally conceived as a 10 ft circle. The design was modified to 9 ft and eventually to 8 ft; the instrument was partly built at each stage and this was one cause of the extraordinary delay. It was completed by Ramsden's successor Matthew Berge (*fl.*1805–51) in 1809.

As with the Palermo instrument, four pillars enclosed a double circle which was carried by a series of radiating cones. At the top of the vertical axis the structure of pillars was held by two great masonary piers, and the base of the axis was also set on a stone pier founded independently of the floor. Four micrometer microscopes were provided for reading the circle.

With his great circle still unfinished, Ramsden died in 1800. Together with Graham, he had been a second major influence on instrumentation in the eighteenth century, and it is appropriate that his late work centred around the development of the circle and so looked forward to the next generation of astronomical instruments.

8

The Longitude Found

THE DEVELOPMENT OF NAVIGATIONAL techniques in the eighteenth century culminated in a solution – or, rather, two solutions – to the longitude problem, and the history of the instruments of the period is bound up with this story, even if they were eventually applied to other purposes, such as improving latitude determination. This major research goal was officially sanctioned and encouraged in England by the establishment of the Board of Longitude in 1714 and the promise of a very substantial reward. By 1716, a large reward was also on offer in France.

The eighteenth-century situation is curious, for the methods rendered possible were neither invented nor truly applied during the period. On the one hand, they had been conceived long before; on the other, their widespread use was not possible until relatively inexpensive and commonplace instruments had replaced the rare and exotic creations of a few leading makers, and until easily learnt procedures had been distilled from impossibly complex mathematical theory. All of this had to await the nineteenth century.

The two longitude methods – lunar distance and chronometer – attracted different sets of supporters and clientele. To the astronomers, and with them the mathematical instrument-makers, the lunar distance method seemed soundly based on astronomical and mathematical principles. They were inclined to regard a chronometer as a mere 'box of tricks', whose principles could not be given a mathematical formulation and whose success depended heavily on the skills of individual craftsmen. Nor were the leading clockmakers adept at explaining their fabulous machines. The protracted dispute, towards the end of the century, over the award of the Board of Longitude's prize, derived in part from the fact that the technical men on the Board were astronomers and mathematicians and the reward was claimed for making a chronometer.

The octant

By the early eighteenth century, the backstaff was a fairly familiar instrument for finding latitude, though the cross-staff continued to be more commonly used by continental sailors. One or two variations are known. The London maker Benjamin Cole (1695–1755)

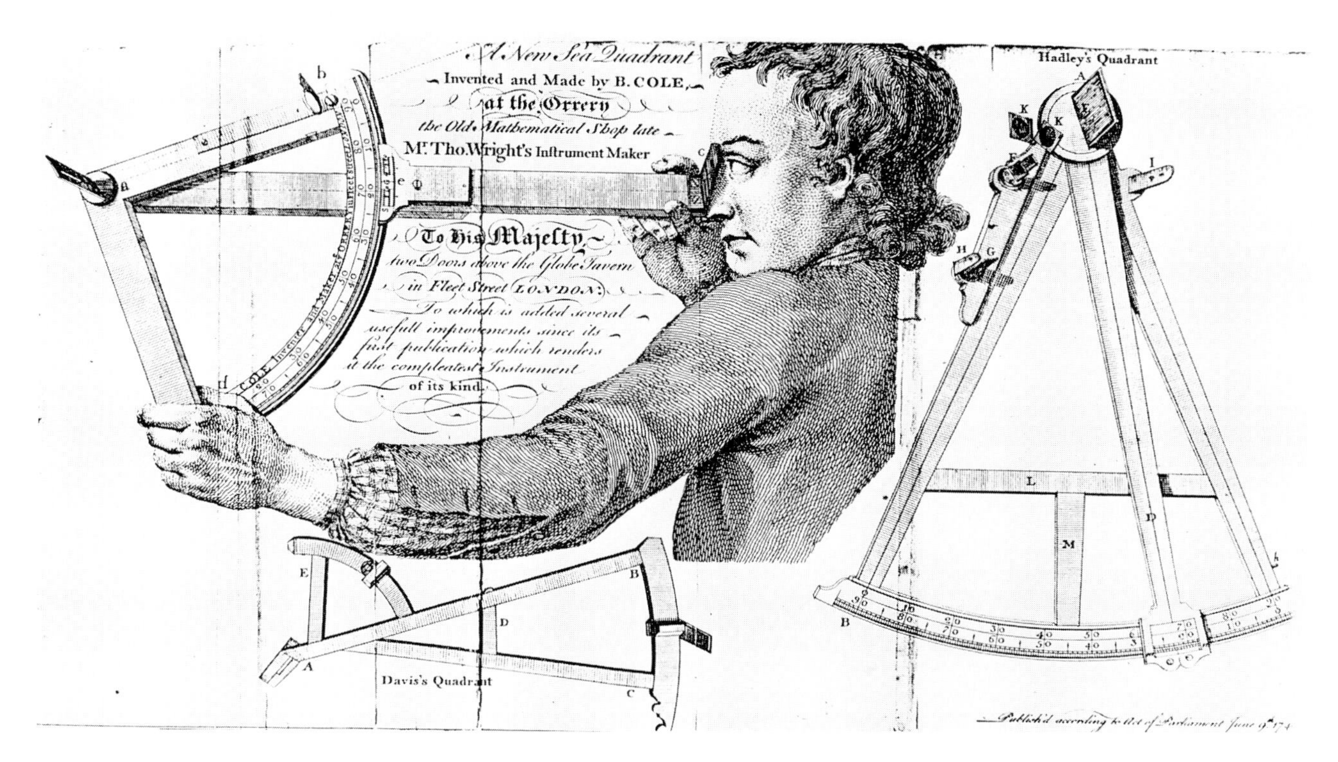

designed a form of back-observation quadrant (fig. 129) that was straightforward to use. An arm of some 2 ft carried – successively – a pin-hole sight, a vernier scale and a horizontal slit. A quadrant, with a shadow vane or lens on its upper radius, pivoted at the horizontal slit and moved across the vernier. The observer simply sighted the horizon and moved the quadrant until the shadow or spot of light fell across the slit. This is an early instance of the vernier applied to a navigational instrument; the backstaff generally had a transversal scale.

'Elton's quadrant' (fig. 130) was closer in form to the backstaff. Designed by John Elton in about 1730, it could be used when the horizon was not visible, as it incorporated several bubble levels. The 60–degree arc was replaced by a straight strut, to which the vane or lens could be pinned at one of three positions, for the three 30–degree ranges of angles to 90°. An index arm, pivoting at the horizon vane,

129 Benjamin Cole's 'sea quadrant' – a form of backstaff – illustrated in relation to the backstaff or Davis quadrant and the Hadley quadrant or octant. The Cole instrument is shown divided to 20minutes of arc and read by vernier to 1minute. (Cole, 1748)

130 Elton quadrant by Sisson, *c.*1732, length 660mm, with lignum vitae frame and boxwood arc. A longitudinal bubble level – a kind of 'artificial horizon' for making observations when the horizon is obscured – and a lamp for illuminating the levels are missing. Greenwich, National Maritime Museum.

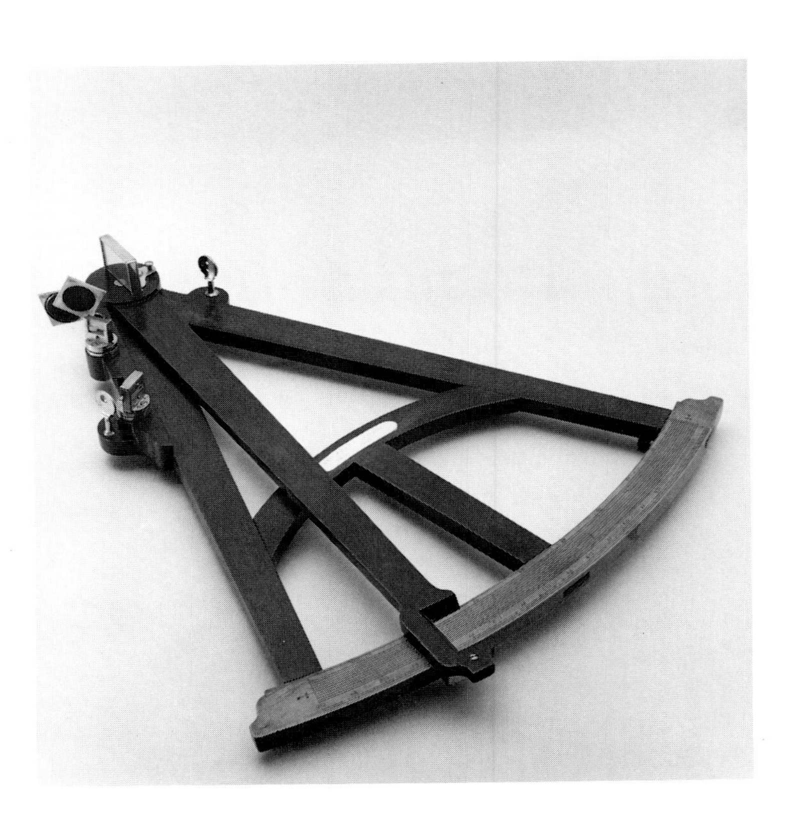

131 (*left*) Octant by Adams, 1753, radius 508mm, with mahogany T frame and boxwood arc. The scale is divided by diagonals, and the back-sight and back horizon glass are clearly shown. Greenwich, National Maritime Museum.

132 (*right*) Octant by B. Messer of London, *c.*1790, with mahogany frame and inlaid ivory scale; the scale is divided to 20 minutes of arc and read by vernier to 1minute. A screw-fit ivory pencil cap is visible at the top of the T brace, which has a central hole to house a pencil, used to note the measurement on a small ivory plate in the back of the frame. The vacant slot between the horizon and back horizon glasses is for the mount for the shades or filters, when they are moved down for the back horizon observation. The adjustment for the index arm is visible, and while the brass arm represents a development, a tangent screw is not yet a standard fitting. Cambridge, Whipple Museum.

moved over a 30–degree arc, with a vernier to give readings to one minute. Both of these instruments are exceedingly rare.

All altitude-measuring instruments were superseded by the octant (figs 131, 132). The idea of using a mirror to bring one target object into coincidence with another, and then noting the inclination of the mirror (half the angle between the objects), was not new, and had been incorporated into instruments proposed by Hooke and by Newton. It is, therefore, not surprising that several suggestions along these lines should appear at about the same time, most notably from Thomas Godfrey (1704–49) in Philadelphia and John Hadley (1682–1744) in London. Other instruments were proposed by the French astronomer Jean-Paul Grandjean de Fouchy (1707–88) and the English insurance broker Caleb Smith (*fl.*1734–45). Hadley originally believed that the octant might be applied to finding longitude by lunar distances, and he proposed two designs to the Royal Society in 1731. An improved design of 1734 became the standard form.

In spite of Hadley's hopes, the octant was almost always used in the vertical plane to measure the altitude of the Sun or a star above the horizon. The horizon is viewed through a pin-hole (later telescopic) sight and the clear portion of a half-silvered mirror, known as the horizon glass. A second mirror, the 'index mirror', is mounted at the pivot of an arm that moves across a scale of degrees. The observer adjusts the index arm until he sees the body whose altitude is to be measured, after reflection in both mirrors, in coincidence with the horizon viewed directly, or to use the common expression, he 'brings the body down to the horizon'. The angle between the two mirrors is measured by the scale and is half the required altitude; for this reason, a scale covering 90° is engraved on an arc of 45°, and the instrument is known alternatively as Hadley's quadrant or octant.

In addition to an improvement in accuracy, an important practical advantage of the octant over the backstaff was that the coincidence of Sun or star and horizon was not affected by the motion of the ship. Thus octants began to be made commercially from the mid-eighteenth century. In theory, these should have been superseded in turn by the more accurate and more versatile sextant, but in fact, because they were cheaper, they continued to be made and used throughout the nineteenth century.

The earliest instruments were made in mahogany and had diagonal scales on boxwood, either inlaid or attached as a separate limb. The fittings – pin-hole sight, mounts for mirrors and shades, clamping screw for the index arm – were of brass, and an inlaid ivory plate carried the maker's inscription or the owner's name, sometimes both. At first, for measuring the Sun's altitude, two alternative shades (red and green) were fitted on a hinge below the index mirror, but three soon became more common.

Gradually, the large mahogany frames were replaced by smaller frames of ebony, the change being more or less complete by the end of the eighteenth century. The index arm also became more commonly made of brass rather than wood. Some octants were made entirely of brass, but these were more common in the later nineteenth century. An important change took place in the scale, with inlaid ivory scales, read by a vernier on the index arm, taking over from the diagonal scales on boxwood. The vernier was already more common from about 1770, though the diagonal survived for a time on cheaper instruments.

The type of vernier fitted can be useful in dating an octant. Until about 1780 a central zero was common, thereafter the familiar vernier with the zero on the right. A clamping screw was used to set the index arm when the sight was completed, so that the scale could be read at leisure. A tangent screw was later added – at first in only the finest and most expensive instruments – for making the final adjustment with the arm clamped.

A particular advantage of the octant was that it could be adjusted for instrumental error. When the index and horizon mirrors are parallel – that is, when the direct and reflected images of a distant object are in coincidence – the index should read zero on the scale. The index mirror is therefore made adjustable by clamping screws, and the horizon glass is often also adjustable by a lever. Both must also be perpendicular to the frame, and this too can be corrected.

Most eighteenth-century octants, and many of the early nineteenth century, have a second pin-hole sight and a second horizon glass, set together on the frame below the normal horizon glass. These are known as the 'back-sight' and 'back horizon glass', and were intended for use when measuring angles greater than 90°. The term 'back-sight' signifies that the observer then had his back to the object and was measuring its angular distance from the far horizon. The back horizon glass should be at exactly 90° from the index glass with the index at zero, and had the effect of adding 90° to the angle indicated on the scale. It was hoped that this would be useful in taking lunar distances, where large angles can be required, but in practice it was difficult to adjust the back horizon glass to its proper position, and it was little used.

Unlike the sextant, the general shape of the octant, with a few exceptions, changed hardly at all. The T frame – a reference to the shape of the central brace – was almost universally adopted. A feature of many instruments was a hole at the top of the T, intended for a pencil, which was set in a threaded ivory cap. This was usually combined with an inlaid ivory note-plate on the reverse of the frame, used for noting the observation.

The reflecting circle

A major problem with the lunar distance method for longitude was the lack of a precision instrument suitable for use at sea. The octant at last approached the order of accuracy required, and a second problem was also nearing solution. The observatories had now provided the empirical basis for an adequate lunar theory. The theory itself was initially the work of the German astronomer Johann Tobias Mayer (1723–62), working in Göttingen at around the middle of the century. Mayer was aware of the reward offered by the British Government, and his first communication with the Admiralty in 1755 initiated a protracted negotiation with the Board of Longitude. It was not until after his death that the British Parliament granted Mayer's heirs £3,000.

Mayer's lunar tables provided the basis for the lunar distance method, published in 1763 by the Cambridge astronomer Nevil Maskelyne (1732–1811) in *The British mariner's guide*. After Maskelyne was appointed Astronomer Royal, they also became the basis of the lunar distance tables in the new annual ephemeris *The nautical almanac*, first prepared at Greenwich for the year 1767. Although this was the first regular provision of lunar distance tables, the *Nautical almanac* was not the earliest nautical ephemeris. A number of sporadic publications, of differing value, had grown from the enterprise of various private individuals, and from 1678/9 the Paris Observatory had published the *Connoissance des temps*. In 1798 this was taken over by the Bureau des Longitudes, the French equivalent of the Board of Longitude. The *Connoissance* included lunar tables, establishments (information for calculating the times of tides at different ports), tables of eclipses of Jupiter's satellites, and so on.

Mayer also designed an instrument based on the same principle as the octant, but with a full circular limb divided into 720°. His account illustrates an index arm and a telescope, both pivoted centrally. In use, the index arm is first set to zero, and the telescope rotated until the two images of a star are seen in coincidence – the one directly, the other by double reflection. The mirrors are then parallel. Now with the instrument in the plane of the angle to be measured and one object viewed directly through the telescope, the index arm is freed, and rotated until the other object is seen in coincidence after double reflection. The angle has now been found, but the double operation is repeated a number of times, and the final

133 Borda-type reflecting circle by Secretan, mid-19th century, radius 135mm, with the telescope, horizon glass and shades on one arm, and a separately pivoted index arm and mirror. Clamp and tangent screws to both arms are used in turn when 'repeating' the observation. The means of adjusting the height of the telescope above the circle can be seen, and is essential to sighting both the direct and reflected rays. Cambridge, Whipple Museum.

angle divided by this number to find a mean value. For this reason, the instrument was sometimes called a 'repeating circle'.

An improved form of repeating circle (the instrument might be called a 'reflecting circle', or even a 'repeating reflecting circle') was designed by the French physicist and mathematician Jean-Charles Borda (1733–99). His *Description et usage du cercle de réflexion* was first published in 1787 and, unlike the Mayer instrument, the Borda circle (fig. 133) and derivative versions were made in some numbers. In his design the arm carrying the telescope was extended right across the circle; the telescope and a clamp and tangent screw were at one end, and the half-silvered horizon glass at the far end from the eye.

The method of using a Borda-type circle is as follows. With the index arm clamped, the observer first sights directly on the right hand object and by reflection on the left, moving the telescope arm until this is achieved. He then frees the index arm, sights directly on the left hand object with the telescope arm clamped, and moves the index arm until the two are again in coincidence. The difference in the readings of the index arm is, in this case, twice the angle required, so that the final aggregate reading must be divided by twice the number of double operations. This represents a saving of time over the Mayer circle.

Borda-type circles were most commonly made in France (by, for example, Jecker, and later Gambey and Secretan), but English examples are known (for example, by Dollond, and in the nineteenth century by James Allan).

The more common form of English reflecting circle, however, followed the design of Edward Troughton (fig. 106), which dates from the last years of the eighteenth century. This was a simple reflecting circle – not a repeating circle – with a fixed telescope; in order to reduce the error involved in defining coincidence between the images of two objects, several observations had to be averaged. Two other sources of error were, however, dealt with by providing a triple index, in the form of three arms, 120° apart and each with a vernier. This reduced the errors associated with reading the scale and with eccentricity. A Troughton circle is one of the most handsome of navigation instruments. The limb is supported by a lattice framework in brass. There is a double mahogany handle and, if the instrument was used by an astronomer, possibly also a brass stand. Scales and verniers are generally on inlaid silver, occasionally on gold.

The sextant

The Board of Longitude commissioned a Mayer circle from Bird, and organized trials. These were carried out between 1757 and 1759 by Captain John Campbell (*c.*1720–90) of the Royal Navy, who found the circle cumbersome to use at sea, but appreciated the value of a range of angles larger than the octant. He thus evolved a compromise design, where the octant arc was enlarged to 60° (for measuring angles up to 120°), and a telescopic sight, which had been suggested by Hadley, was reintroduced. He had the first 'sextant' made by Bird.

From the beginning, sextants were generally made in brass, and early examples were large and heavy. An example made by Bird in about 1770, in the National Maritime Museum, London, is fitted with a pole for supporting the instrument by a belt worn by the observer. This problem was effectively solved by one of the most crucial developments in the history of mathematical instrument-making – the mechanical division of scales, which dates from the dividing engine built by Ramsden in 1775. This allowed much smaller arcs to be divided to the required accuracy.

The sextant was essentially a precision instrument for measuring lunar distances. It could, of course, also be used to find latitude, but it became customary for the seaman to use an octant for altitude

134 (*left*) Sextant by Ramsden, *c.*1785, radius 300mm, divided to 20 minutes of arc and read by vernier to 30 seconds. The lattice frame has four struts, jointed together where they cross. The clamp and tangent screws and the central reinforcement on the index arm became standard features of sextants and octants. Cambridge, Whipple Museum.

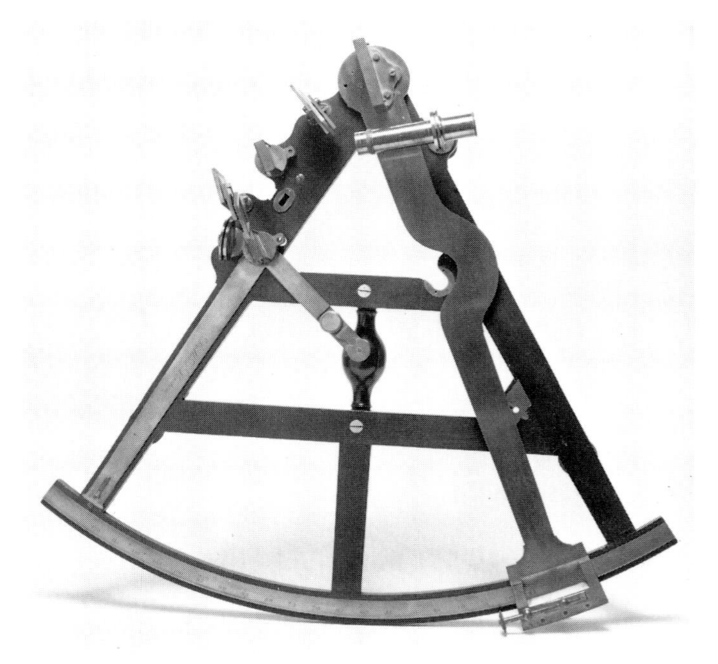

135 (*right*) Sextant by Dollond, *c.*1772, radius 450mm, of a special design by Peter Dollond to facilitate adjustment of the back horizon glass. This should be at right angles to the index mirror when the index reads zero, thus adding 90° to the angle indicated by the scale. In Dollond's design the back horizon glass is carried at the pivot of an arm which rotates through 90°, when the glass should be parallel to the index mirror, and with the index at zero the two images of a distant object should coincide. The index arm is kinked to allow it to be set at zero with the telescope moved into position for this adjustment. Cambridge, Whipple Museum.

136 Sextant by Ramsden, *c.*1795, radius 165mm, with a silver scale, read by vernier to 20 seconds of arc. This is the 'bridge' frame: the upper half is double and encloses the optics (cf. fig. 202). The instrument is stowed in the 'keystone' box, which also acts as a base for the pillar stand. Ramsden's sextants have serial numbers; this one is '1209'. Cambridge, Whipple Museum.

measurements. Thus the octant took on the role of the familiar, workaday instrument, while the sextant was more special, to be treated with great care and used only when precision was essential. Every feature that might enhance accuracy was considered – a rigid frame, a carefully divided arc, a vernier scale, a tangent screw and telescopic sights. Before long, a reading glass or microscope would be added.

Most thought was given to the frame, in an effort to combine rigidity with lightness. Ramsden himself made several types. One had a lattice structure (fig. 134), with brass struts crossing diagonally between the radii and the limb. Another Ramsden form – the 'bridge type' (figs. 136, 202) – had an open brass framework built over the upper half of the frame, which both reinforced the structure and protected the optics. One of the most successful eighteenth-century designs was that patented by Edward Troughton in 1788 – his 'double-frame' or 'pillar-frame' sextant (figs. 137, 203), in which two thin T frames of plate brass were held together by a series of brass pillars. According to Pearson, such sextants were 'in great request in the

137 Sextant by Troughton, *c.*1790, radius 250mm, divided to 10 minutes of arc and read by vernier to 10 seconds. An example of the 'double' frame, mounted on a pillar and tribrach stand, with two counter-weights, which allow the sextant to be adjusted to any position. The serial number on this instrument is '200'. Cambridge, Whipple Museum.

naval service' (Pearson, 1828, p.576), and they were made throughout the nineteenth century. Troughton applied the same double-frame principle to a design of a portable astronomical quadrant.

Mechanical division

While there had been earlier attempts at the mechanical division of scales, the first real impact on the manufacture of instruments was due to Ramsden. The primary components of his dividing engine (fig. 138) were a racked circle engaging an endless screw. The heavy circle, cast in a brass-type alloy known as bell-metal, was 54 in in diameter and had ten radial pieces, reinforced by a perpendicular ring and braces. The rim was cut into 2,160 (360 × 6) teeth. Thus one revolution of the steel screw turned the wheel by ten minutes. A circular head attached to the screw was divided into sixty parts, each one representing a movement by the great wheel of ten seconds.

The whole was mounted on a stout, cross-braced tripod stand in mahogany. An instrument's frame could be attached to the wheel, while above a carriage for the dividing point ran on a frame between the centre and the mount for the screw, and could be clamped at the appropriate radius. Thus the point was fixed while the wheel, with the scale to be divided, was moved beneath it by the action of a foot treadle. A number of sophistications were built into the design, and all were described in Ramsden's published account. He also explained his method of cutting the teeth of the wheel and the machine for making the screw, on which the accuracy of the division would entirely depend.

The Board of Longitude were anxious that Ramsden's work should have the maximum possible impact on instrument manufacture, and it was due to their efforts that this was so. They granted him a reward of £300, and for a further £315, purchased his engine (though it remained in Ramsden's hands) and his agreement to certain conditions. He was to submit a full written explanation and description of the engine, instruct a number of other makers in its manufacture and use, and undertake to divide instruments brought to him by other makers on terms set by the Board – 'at the rate of Three Shillings for each Octant, and at the rate of Six Shillings for each brass Sextant' (Ramsden, 1777, preface). The Board also published Ramsden's account and drawings.

138 Ramsden's dividing engine, also illustrated in fig. 88. (Ramsden, 1777)

Before long, dividing engines were made in England by John Troughton and Edward Troughton, and in France by Lenoir, Fortin and Jecker, among others. It was a development of the greatest importance for the trade. Not only could instruments be divided with greater accuracy, and thus made smaller, but the process was also speeded up enormously. Hand division was a tedious and exacting task, and a skill on which a maker – such as John Bird – could found a reputation. Now a crucial *ad hominem* element had been removed; no longer could a maker be celebrated for his dividing skill, though some made their names by designing improved engines. It was an early instance in the instrument-making trade of the new industrial methods.

The chronometer

The early history of the chronometer most commonly brings to mind – at least in England – the figure of John Harrison (*c.*1683–1776), and it is generally a romantic picture of a relatively uneducated Yorkshire carpenter and his struggle for recognition and reward with the bureaucratic machine. He combined great ingenuity with great perseverance, and his story has produced a large literature to which we cannot do justice here. Best known among his inventions, perhaps, are the gridiron pendulum and the grasshopper escapement. In the former, a combination of brass and steel rods, arranged to expand in opposite directions in heat, were used to overcome the effects of temperature change on the length of a metal pendulum. Harrison's escapement – the 'grasshopper' – acted with no friction between the pallets and the escape wheel, and so needed no oil. Other innovations were a remontoire (a device for isolating the escape wheel from variations in the driving force transmitted through the train), and a form of maintaining power (a means of ensuring that the driving force is maintained and the timepiece does not stop while it is being wound).

In his quest for the longitude prize, Harrison's first marine timekeeper (fig. 139) was completed in 1735. It stood some 40 cm high, and its going was regulated by the counteracting motions of two bar balances, pivoted centrally and controlled by four helical balance-springs. The movement incorporated two modified grasshoppers, a gridiron to vary the tension of the springs as their elasticity changed with temperature, and Harrison's maintaining power. The whole machine was slung in gimbals and carried by a spring-mounted frame.

Sea trials, organized by the Board of Longitude, were encouraging, so that Harrison's second marine clock followed much the same lines, save for the inclusion of a remontoire, and was completed in 1739. His third was not finished until 1757. Though built on a scale similar to the earlier clocks, the third was a far more complicated and sophisticated machine. It would be wrong to pretend to describe it here, save perhaps to say that the length of the spiral balance-spring that controlled the motions of two large balance-wheels (replacing the former bar-balances) was adjusted automatically for temperature by a 'compensation curb' – a bimetallic strip of brass and steel.

Harrison's fourth timekeeper, the prizewinner, looked quite different from the previous three. A large silver-cased pocket-watch, it incorporated the compensation curb and the form of remontoire used

139 Harrison's first marine chronometer, 1735, height 408mm. Greenwich, National Maritime Museum.

in the third clock, but not, on this occasion, the grasshopper. After successful trials on two voyages to the West Indies and an epic struggle between Harrison and the Board, the watch eventually won the £20, 000 prize. One feature of the struggle was that Harrison felt that Maskelyne, *ex officio* an influential member of the Board, was an unfair judge because of his commitment to lunars.

Harrison's success naturally encouraged other makers in England, such as Larcum Kendall (1721–95) and Thomas Mudge (1715–94), but the historical antecedents of the standard marine chronometer are really found in France. Harrison's work, though ingenious and attracting interest in the chronometer method, did not embody the principles that made a success of the common chronometer of the nineteenth and twentieth centuries.

In France, the chronometer was pioneered by Pierre Le Roy (1717–85) in competition with the Swiss-born Ferdinand Berthoud (1729–1807). Harrison's timekeepers represented, in truth, a dead-end as far as the development of horology was concerned. Le Roy, on the other hand, sought out new techniques which, with modifications, formed the basis of the successful working chronometer. The essential principles of the chronometer are the detached escapement and the compensation balance. The former is a device for giving impulse to the balance (in order to maintain its motion) only for an instant in its oscillation cycle, the escape-wheel being locked at other times. The balance then moves with minimum interference from the escapement. The compensation balance has a built-in adjustment to its inertia, which counteracts the changing elasticity of the balance-spring with changing temperature. Le Roy achieved this in his famous '*montre marine*' of 1766, by forming the balance of mercury thermometers; he also devised the more common method of forming the balance arms of bimetallic strips.

In England, and in France, masterpieces of horology had been made capable of finding longitude at sea, but this did not represent a solution to the longitude problem. A useful navigational instrument had to be capable of manufacture in bulk, and by much less expensive methods. The achievement of fashioning a serviceable and reasonably priced instrument belongs largely to two English makers, John Arnold (1736–99) and Thomas Earnshaw (1749–1829). Although the process began in the eighteenth century, the standard marine chronometer is typically a nineteenth-century instrument, and is discussed in Chapter 12.

140 Mechanical log as patented by William Foxon, *c.*1772, which registered the rotation of a fly towed behind the ship in terms of distance travelled. Alternatively the dial 'By One Minute Glass' gave the speed in knots; the glass was turned at zero and the reading noted after one minute. London, Science Museum.

Marine compasses

Despite the progress made in the sextant and the chronometer, surprisingly little improvement was made in the compass, an instrument vital to the safety of a ship and her crew. There was a general shift from compass bowls turned in wood to those cast in brass, but in America, for example, wooden bowls were still common in the nineteenth century, and while the British Navy had declared brass bowls standard in the eighteenth century, they were not universally adopted, and pewter or copper bowls have been found, as well as

141 Azimuth compass by Rust & Eyre, *c.*1770, diameter 187mm, gimbal mounted in a wooden box or binnacle, and fitted with slit-and-window sights. Greenwich, National Maritime Museum.

ones in wood. The French, after some early experiments with brass, were using square wooden 'bowls' or boxes in the eighteenth century.

Compass cards were also being altered, but not very systematically. Early examples were coloured, often in red and blue (blue for the four cardinal points, red for the half-cardinal), and cards decorated with ships, animals or allegorical figures, for example, are found among eighteenth-century Danish compasses. By the end of the eighteenth century, however, black and white cards were more common – their markings were seen more easily at night – though designs still varied greatly in detail. 'Tell-tale' or 'hanging' compasses, intended to be read from beneath – they might, for example, have been hung over the captain's bunk – had cards with east on the left and west on the right.

With regard to the magnet itself, different shapes and combinations of needles were tried in early dry-card compasses. One common early form had two wires, each bent once, set in a diamond shape, or a single wire bent twice with the same result. While two such magnets might easily achieve a mechanical equilibrium, magnetic balance between the two was more difficult. An alternative was to use a single flat needle with a central aperture, but changes were made on an *ad hoc* basis; there were no underlying principles.

The earliest systematic attempt to improve the marine compass occurred around the middle of the eighteenth century, through the work of the London physician Gowan Knight (1713–72). He had developed a method of strongly magnetizing needles and bars, by successively combining and reinforcing a number of bar magnets. His reputation attracted the attention of the Admiralty and led to his examining compasses in use in the Royal Navy. He then submitted designs to replace the ill-informed methods and haphazard procedures he had discovered. Among his innovations were a single straight needle, in the form of a steel bar, a replaceable pivot and an agate bearing in the cap. Although not all his ideas were adopted, his work was important in focusing attention on the compass and revealing its inadequacies. Compasses after Knight's design were made by George Adams and issued to Royal Navy ships, but examples are now exceedingly rare.

Navigation in the nineteenth century was to benefit from a thorough re-examination of the marine compass, for which Knight's work had at least prepared some of the ground. The two longitude methods that were made possible in the eighteenth century also remained to be fully applied in practice, through the provision of large numbers of reasonably priced instruments and the dissemination of the techniques involved in their use. The methods used by many ordinary seamen – with the exception, perhaps, of the use of the octant – had not yet changed substantially. But finding the longitude had created the promise of a change as fundamental as that from the plane to the Mercator chart, or from bearing and distance to latitude sailing.

9

Geodetic and Commonplace Surveying

IGHTEENTH-CENTURY SURVEYING INstruments fall into two, almost separate, classes. We have already noted the large-scale, officially sponsored, geodetic exercises, embarked on first in France, and taken up later in the century in Britain. The finest instruments were commissioned for this kind of work, equipped from the beginning of the period with telescopic sights and other refinements derived from the instruments of astronomy. The ordinary surveyor, engaged on an estate survey or on drainage work, had neither the funds to purchase a sophisticated instrument, nor the inclination to master its complexities.

For everyday surveying, the basic equipment of a plane table and a chain, perhaps a circumferentor and, at the very most, a simple theodolite, were more than sufficient, and these instruments represent the common tools of surveying throughout the century. From time to time makers sought to promote greater sophistication – new levels or, notably, altazimuth theodolites – but with only modest success. A maker might expect to benefit financially from a shift in the instrumentation recognized as necessary to the surveyor's art. But it would not be easy to change the practice of surveyors, and to introduce them to the developments taking place elsewhere in practical mathematics.

Geodetic instruments

We have seen that the early French expeditions were equipped with portable quadrants, and occasionally with portable transit instruments and zenith sectors. Octants were also sometimes carried: one English example was taken on La Condamine's expedition to Peru as early as 1735. Later in the century, a form of repeating circle, whose design was probably due to Lenoir, was used in the triangulation link between Paris and Greenwich. Two telescopes were pivoted on either side of the circle, and the whole given an altazimuth stand. The Borda repeating circle – which, of course, had a single telescope and worked by reflection – was also used in French geodesy, and examples were made by Lenoir, with stands for terrestrial use.

On the English side, the beginning of serious geodetic work was marked by Ramsden's first great theodolite (fig. 142) of 1787,

142 Ramsden's first great 3 ft (920mm) theodolite, 1787, commissioned by the Royal Society for use in the Greenwich to Paris triangulation. (Mudge, 1799)

purchased by the Royal Society for use by Roy on the Greenwich-Paris link. The 3 ft azimuth circle was attached to the vertical column by radial cones (of which there were ten), in a manner similar to the lower circles on the Palermo altazimuth and the Shuckburgh equatorial. The circle was divided to fifteen minutes, could be read to a second of arc by two micrometer microscopes, and estimated to a fraction of a second. The vertical column tapered to a yoke for the horizontal axis – composed of two opposed cones – supported at either end by tapering ladders rising from the column. The 3 ft reversible telescope had a micrometer for small vertical angles, which could otherwise be taken from a 6 in semicircle, divided to half a degree and read by a micrometer microscope to five seconds. A lantern was used to illuminate the cross-wires, by light introduced into the hollow axis and then reflected down the tube. Both telescope and axis had hanging levels.

Ramsden's 'portable' zenith sector was the other great instrument in this period in England. Its 8 ft telescope moved across an arc divided to five minutes, and its position could be estimated by micrometer microscope to 0.1 seconds. Its containing structure was carried in an outer frame 12 ft high on a 6 ft square base.

Ramsden's second 3 ft theodolite (fig. 143) (the one made for the East India Company in 1790, but purchased by the Board of Ordnance) had a complete vertical circle – double, with connecting bars, in the manner of his contemporary astronomical work, but still of only 6 in radius. The 3 ft azimuth circle was now divided to ten

143 Ramsden's second 3 ft (920mm) theodolite, 1790 Note the four micrometer microscopes for reading the horizontal circle, its rack-and-pinion adjustment and the cones supporting it. London, Science Museum.

minutes and read by four micrometer microscopes. This emphasis on azimuths was appropriate to the principal instrument of a large-scale triangulation, but also parallels the everyday surveying tools of the same period.

Theodolites

Throughout the eighteenth century, the word 'theodolite', used without further qualification, referred to an instrument for measuring azimuths only – to what we have called a simple theodolite. While the graphometer remained popular in France, in England the simple theodolite was not generally displaced by attempts to introduce the altazimuth instrument.

Bion's *Traité* of 1709 described a '*planchette*', for use in triangulation, as a divided circle with a centrally pivoted telescopic sight, connected to a tripod by a ball-and-socket joint. He also described the graphometer and a quadrant for taking both altitudes and azimuths.

In 1723 Stone translated Bion's *planchette* as 'theodolite'. His section on English instruments described, in addition, a 'theodolite' with the telescope slung underneath the circle – able to pivot about a horizontal axis, so as to take in objects at different levels, but without a vertical arc. He added:

> Note, There are some Theodolites that have no Telescopes, but only 4 Perpendicular Sights; two being fastened upon the Limb, and two upon the ends of the Index. (Bion, 1723, p.127)

In fact, this instrument became much the more popular. It was a divided circle (fig. 145) with a pair of plain sights, diametrically opposed and used to set and check the instrument's alignment, and a pair of sights on a centrally pivoted index, used to take bearings from the circle. The pivoted index was often equipped with a compass, itself with a divided circle, and in this case the index plus compass could function, in effect, as a circumferentor. Bion goes on to point this out:

144 An unusual 'graphometer', early 18th century, diameter 325mm, which in some respects is closer to a simple theodolite, and well illustrates the relation of both to the circumferentor. The circle is continued sufficiently far to allow the alidade to accommodate a sizeable central compass with a silvered rose and circle divided to degrees. It was more usual (fig. 68) to place the compass within the semicircle, and use the fixed sights for the circumferentor function. Here the alidade doubles as a circumferentor of the familiar type. For the theodolite function either end of the alidade may be read, and the semicircle is numbered in both directions, 0–180 and 180–0. Cambridge, Whipple Museum.

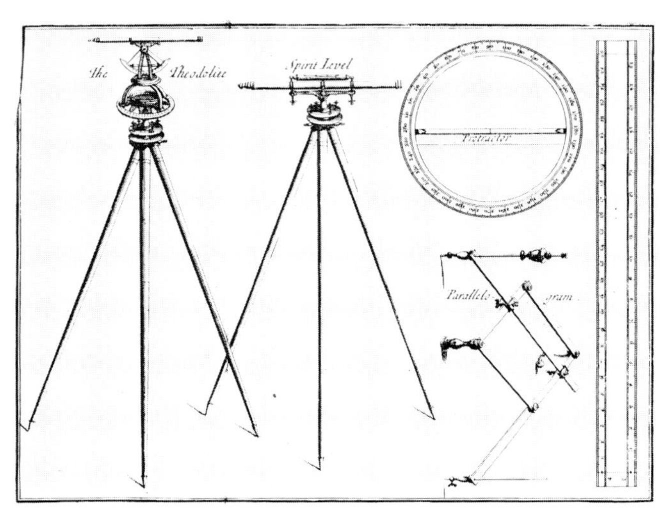

145 (*left*) Simple theodolite, by Pieter & Adrianus Baijens of Delft, early 19th century, diameter 267mm. The instrument can also be used as a cross, or as a circumferentor, and has a ball-and-socket mount and octagonal box.

146 (*right*) Frontispiece to Wyld, 1725, with Sisson's altazimuth theodolite and improved level.

> Note likewise, That the Index, and Box and Needle, or Compass of the Theodolite, will serve for a Circumferentor. (Bion, 1723, p, 127)

This ambiguity may be the source for the confusion in recent terminology, where the whole instrument has been called a circumferentor. But this expression was clearly reserved for a compass with sights, where readings were indicated by the magnetic needle instead of an index arm. The fact that this form of simple theodolite could serve both functions helps to explain its enormous popularity in England. Some examples had an additional pair of fixed sights, at right angles to the first pair, so that the theodolite could act as a surveyor's cross (see below).

This is repeated in Stone's 1758 edition of Bion, but two small paragraphs in the new 'Supplement' acknowledge some serious attempts in the meantime to introduce an altazimuth instrument. They are headed 'Of Mr Sisson's Theodolite':

> This is certainly the best, most complete, handsome, and well designed instrument possible, not only serving as a Theodolite to take horizontal Angles, but likewise to take vertical Angles as a Quadrant . . .
>
> There is likewise another very good Theodolite, made by Mr Heath the Mathematical Instrument Maker, whose uses are to be seen in Hammond's Surveying, wrote in reality by the late very ingenious Mr Samuel Cunn (who being a Butcher, that kept a Butcher's Shop in Newport Market) was also a very great Mathematician: One of the best Measurers of Artificers Works, Surveyors of Land, and Expounders of Euclid, and Apollonius, in the World. (Bion, 1758, p.302)

The episode Stone referred to began in 1725 with the appearance of Samuel Wyld's *The practical surveyor*, of which Jonathan Sisson was one of the publishers. Wyld covered the standard instruments – the plane table, the chain, the circumferentor and the simple theodolite – but then described a new theodolite and a new level by Sisson, both of which were prominent on the frontispiece (fig. 146). The theodolite was an altazimuth, with a telescopic sight. From an azimuth circle with a compass, three curved supports rose to a central platform, on which was based a single 'A' frame carrying the pivot for a telescope and an altitude quadrant.

In the same year, a second textbook with the same title was published by the instrument-maker Thomas Heath (*fl.*1714–65). The author named was John Hammond but, as Stone observed, the book was actually written by the butcher Samuel Cunn. An improved theodolite is referred to on the title-page and in the text, but without illustration or detailed description. Heath used subsequent editions of this book to introduce his design of an altazimuth theodolite. The second edition of 1731 has an account in a new appendix and an illustration on a frontispiece (fig. 147). By 1750, the description is included in the text, and there are explanations of two surveys, 'the one performed by the common [simple] Theodolite, the other by the new improved Theodolite' (Hammond, 1750, pp. i-ii). On Heath's instrument, the vertical arc was a full semicircle, but with its curved side uppermost (we shall refer to such an arc as 'inverted') and racked, so that the telescope, pivoted at the centre of the semicircle, was moved along the arc by a pinion. The semicircle rose directly from its support on the azimuth circle.

Sisson's rejoinder was to publish a further textbook, *Practical surveying improved* by William Gardiner, in 1737. The theodolite and level, both incorporating further improvements, were again illustrated on a frontispiece and described in the text, which dealt with both the 'common' theodolite and 'Mr Sisson's latest improved Theodolite'. The single A frame now rose from the azimuth plate, without the addition of a curved tripod, and the vertical arc was a semicircle.

In Germany, Brander made several different types of altazimuth theodolite in the second half of the century (fig. 91). The most common form had a telescope attached to the diameter of a vertical semi-

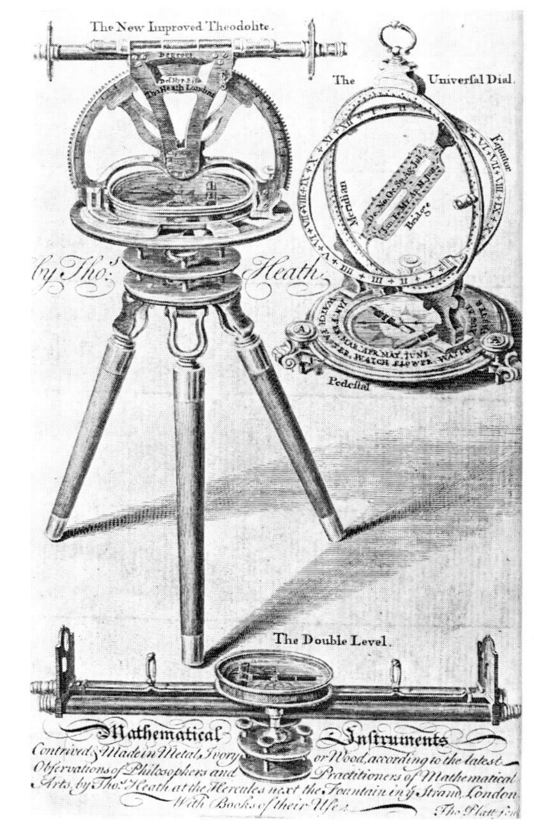

147 (*above*) Frontispiece to Hammond, 1731, showing Heath's altazimuth theodolite (cf. fig. 103), and below a level with two telescopes for sighting in both directions.

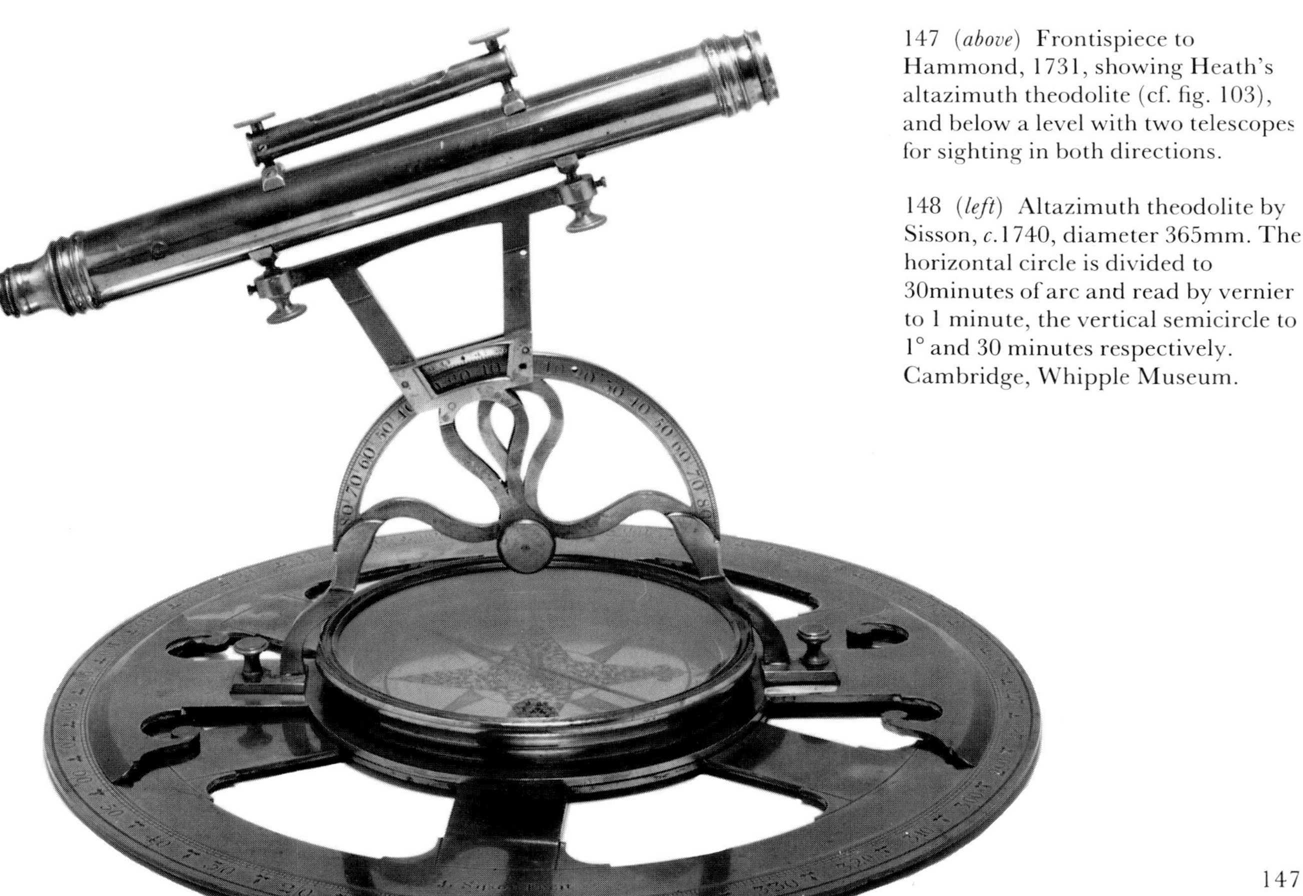

148 (*left*) Altazimuth theodolite by Sisson, *c.*1740, diameter 365mm. The horizontal circle is divided to 30minutes of arc and read by vernier to 1 minute, the vertical semicircle to 1° and 30 minutes respectively. Cambridge, Whipple Museum.

circle, moved by a pinion engaging teeth on the inner edge of the arc. The mount for telescope and arc was simply attached directly to the index arm on the azimuth circle, which must have been less steady than examples where it was carried on a circular plate, moving on or within the divided circle. A more simple model had an inverted arc, a more sophisticated one had an endless screw engaging teeth on the outer rim of a vertical arc, more securely mounted on a vertical column with a projecting index arm.

In spite of the enterprising assaults on surveying tradition made by Sisson and Heath, when George Adams published his standard work, *Geometrical and graphical essays*, in 1791, the 'common' theodolite he described was similar to that noted by Stone in 1723 – an azimuth circle with two fixed sights and a pivoted alidade with a compass. Adams, himself an enterprising maker, naturally went on to describe more sophisticated, altazimuth instruments. The simplest of these had an inverted semicircle, after the manner of Heath's design (fig. 79). Two more complex examples, one of which was ascribed to Ramsden (fig. 152), each had a double A frame on the azimuth circle, supporting the horizontal axis of a racked semicircle. The semicircle carried a telescopic sight, reversible in two Y supports.

Such was Adams's optimism that a still more complicated theodolite was described in the preface and illustrated on the frontispiece (fig. 149), after completion of the text. Here the vertical semicircle was beneath the azimuth circle, after the manner of some of Adams's

149 Theodolite by Ramsden, illustrated on the frontispiece to Adams, 1791.

150 Theodolite by Heath, *c.*1725, with a short altitude arc, and a second telescope pivoted beneath the horizontal circle, for establishing and checking the instrument's orientation. London, Science Museum.

151 A combined simple and altazimuth theodolite by Benjamin Cole, c.1750, diameter 196mm. The vertical arc and telescope on the right can replace the plain sights on the index arm, and the pair of fixed sights can also be removed, as they would interfere with the telescope. This turns a simple theodolite into the more sophisticated altazimuth form. The altitude is given by an index arm; it soon became standard for the vertical arc or circle to move. Cambridge, Whipple Museum.

portable astronomical instruments, and two telescopes were mounted in axes on double A frames carried above the azimuth plate and below the vertical arc. The upper telescope – the one moved into alignment with the target – thus had no direct connection with the vertical arc, but could be clamped and then levelled to discover the required altitude. Such complexities led to the instrument being omitted by the subsequent editor, William Jones.

These are some prominent instances of the makers' enthusiasm for unwanted sophistication. The ideas they promoted made only modest headway with the ordinary surveyor, but the designs they evolved contributed to the standardization of the altazimuth theodolite, finally achieved in the nineteenth century. For the time being, the ever-resourceful makers devised a few compromises. Examples of combined simple and altazimuth theodolites (figs. 103, 151) were made around the middle of the century. In these the ordinary form of simple theodolite was given detachable sights on the index arm; in their place could be fixed a vertical divided arc with a telescopic sight, and the choice was left to the surveyor. Another idea (fig. 150) was to provide a pivoted telescopic sight, whose eye-end could be clamped to, or moved by rack and pinion along, a short vertical arc of some 10° above and below the horizontal. Adams described the latter type of instrument, and it is interesting that both forms of compromise were made by Heath.

152 Plain theodolite illustrated in Adams, 1791, and attributed to Ramsden. Note the bevelled or chamfered edge of the horizontal circle continued in the edge of the moving plate above, which was a feature of Ramsden's work.

Plane table, circumferentor and cross

The plane table retained its popularity among instruments of more general currency. Already in 1723, Stone described the form (fig. 153) that remained standard throughout the century – an oak table on a tripod, with a compass attached in an octagonal box, a boxwood frame or surround divided in degrees, and a brass rule with engraved scales and plain sights. Towards the end of the century, alidades with telescopic sights on vertical arcs were designed and made, but were not commonly used.

The circumferentor also reached its standard form (figs 154, 165) in the early decades of the century, consisting of a magnetic needle pivoted in a circular glazed brass box, engraved inside with a compass rose and a degree scale, and two brass arms with plane sights extending on opposite sides. Occasionally, a plane table alidade might be fitted with a compass, and it could then function as a circumferentor.

Irish and American makers produced significant numbers of circumferentors in the eighteenth century. Adams noted that the instrument was more popular in America than in Britain, and in fact, an entire family of instruments was to develop from the circumferentor in America in the nineteenth century. With the magnetic needle acting as a reference, the circumferentor was appropriate to surveying a new land, uncluttered by artificial landmarks. In Europe, the theodolite or plane table was suited to measuring or reproducing the angles already established on the landscape by fields, houses, churches, and so on.

153 (*left*) Plane table by Benjamin Cole, mid-18th century, showing the scales on the alidade and folding frame. The alidade has a linear scale with transversal division, and others for drawing maps to a variety of scales; the frame has both linear and degree scales. London, Science Museum.

154 (*right*) Circumferentor by Seacome Mason of Dublin, *c.*1790, length 478mm. The detachable sights are fixed by wing nuts, and the silvered compass rose, divided to 1°, has two inset bubble levels. Cambridge, Whipple Museum.

With a plane table or circumferentor, in either case necessarily accompanied by a chain, a complete estate survey was readily achieved. At an even more basic level, a surveyor in whom geometrical ambition was entirely absent, could get by with a cross and chain. The cross came in two forms, open or enclosed. The open form (fig. 155) had four arms, set at right-angles, with four plain sights; the enclosed was a brass box with slits. These were used to establish a straight line, measured by the chain, and 'off-set' lines at right angles, extending to significant landmarks – a tree or the corner of a field.

Even at this level, the maker could not leave the surveyor in peace. By the end of the century, William Jones was describing an enclosed cross (fig. 156) in the form of a cylinder with right-angle sights, moved by rack and pinion on a circular base divided in degrees, and

155 (*right*) Surveyor's cross by J.H. Temple of Boston, *c.*1850, length 279mm, with detachable sights held by knurled screws, and two bubble levels. Cambridge, Whipple Museum.

surmounted by a compass box with a needle and degree scale: 'Thus the surveyor may have a small theodolite [by reading the lower, fixed scale], circumferentor [by noting the needle's position on the upper scale], and cross staff [surveyor's cross] all in one instrument' (Adams, ed. Jones, 3rd edn, 1803, p.199).

The level

As has been noted, the level was one of the oldest instruments of surveying, and early examples were applications of the water level or the plumb-line. It was used to ascertain the level in any direction, and to determine differences in height, by sighting on a stave marked with a vertical scale. Specialized instruments, known as 'drainage levels' (fig. 157), with small vertical arcs, were used to establish slight inclinations. While the water level remained in use throughout the eighteenth century, especially in France, an increasingly popular alternative form had a telescopic sight levelled by a bubble or spirit level.

Sisson's improved level (fig. 146), described in 1725, was of this type. The spirit level was raised on two standards above the telescope, which was attached to a horizontal bar by two adjustable screws, used to set the bubble to the centre of its tube. The bar and telescope rotated above a ball-and-socket joint, whose inclination was controlled by two circular plates linked by four adjusting screws, which passed through the upper plate and rested on the lower. (This form of mount was also used for Sisson's theodolite.) The whole was set on a folding tripod stand.

The level illustrated by Heath in 1731 (fig. 147) had two telescopes mounted in parallel, but in opposite directions, so that fore-sights and back-sights could be taken without moving the instrument. It was also surmounted by a compass. Sisson responded to both these innovations in the developed version of his level (fig. 158), whose illustration and description were communicated to the Royal Society in 1736 and published in Gardiner's textbook the following year. Sisson had, in fact, devised a classic design of level, which became standard – at least in the best instruments – for a century.

The spirit level was now carried beneath the telescope, and one

156 Enclosed surveyor's cross by W. & S. Jones, *c.*1800, diameter 76mm, elaborated into a multi-function instrument. The sights of the cross can be used in conjunction with the divided circle below as a simple theodolite, or with the compass above as a circumferentor. Cambridge, Whipple Museum.

157 (*below left*) Drainage level by Thomas Wright, 1724, length 432mm. The telescope pivots at the object end and its inclination is adjusted over the short arc at the eye end by clamp and tangent screw. Cambridge, Whipple Museum.

158 (*below*) Sisson's Y level, illustrated on the frontispiece of Gardiner, 1737 The upper diagram shows the compass and the bar carrying the Y bearings for the telescope.

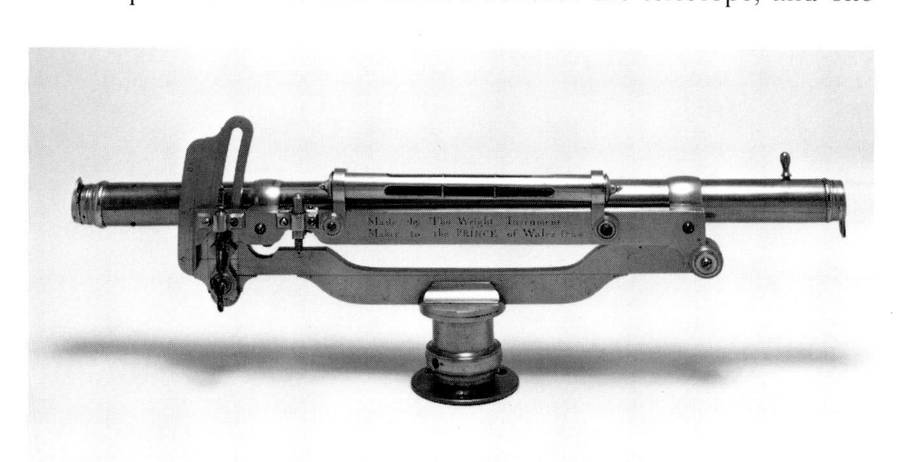

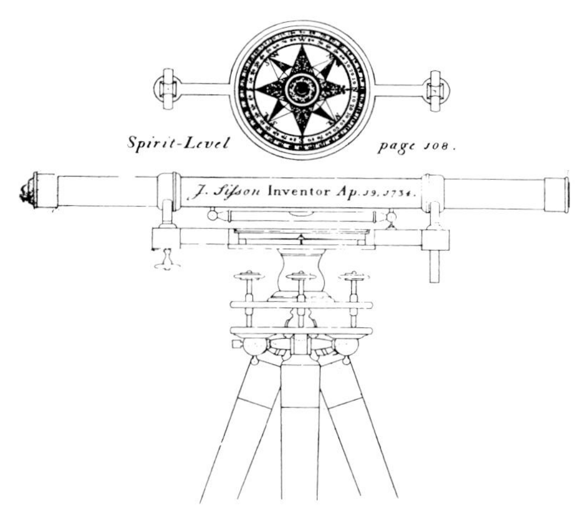

159 Level by Troughton, *c.*1800, length 504mm. The classic Y level, very close to Sisson's original design (fig. 158), but with the bubble mounted above the telescope. The restraining arms above the Y bearings are held closed by detached pins. Cambridge, Whipple Museum.

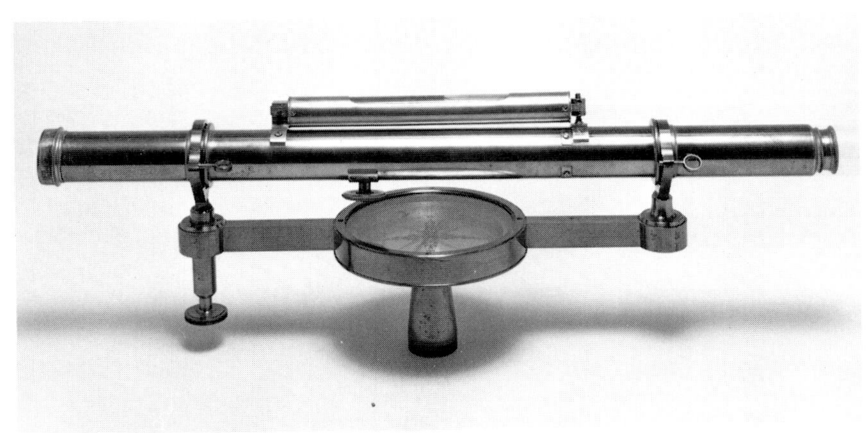

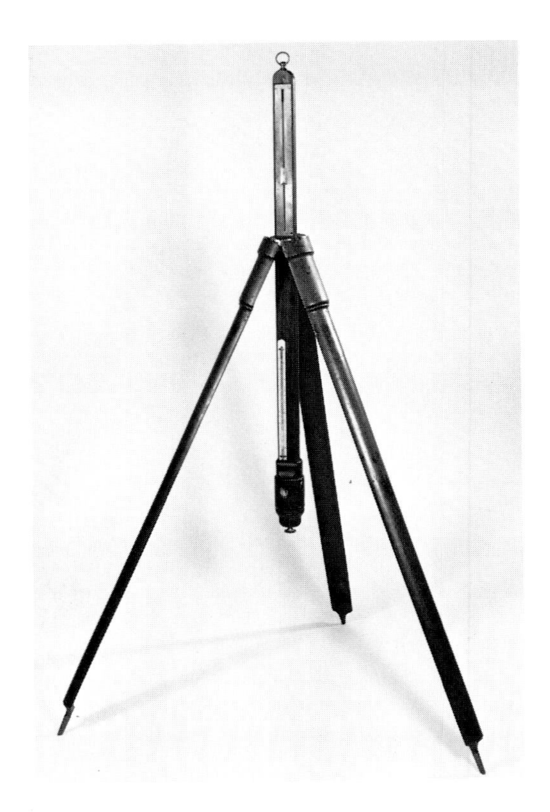

end fitted with a screw to set it parallel. The cylindrical tube of the telescope rested in two supports, 'each nearly shaped like a capital **Y**' (Gardiner, 1737, p.111), and was reversible, so that sights could be taken in opposite directions without the uncertainty of rotating the instrument. Gardiner pointed out that it was for this reason that a level might have two telescopes, but that Sisson's method was more reliable. Two upright supports, as before, carried the telescope above a horizontal bar, but now the height of only one was adjustable by a screw. At its centre, the bar accommodated a compass with a divided circle, in the manner of the circumferentor, so that Gardiner points out that 'it is in all respects a Circumferentor also' (Gardiner, 1737, p.111). The ball-and-socket joint, four adjusting screws and plates, and tripod stand were as before.

The instrument became known as the **Y** level, was described in precisely this form by Adams, and was made well into the nineteenth century (figs 81, 159, 166a). It is rare to find the origin of a surveying instrument dated with any precision, but Sisson had the diagram of his level, published by Gardiner, marked: 'J. Sisson Inventor Ap. 19, 1734'.

160 (*above*) Mountain barometer by Dollond, *c.*1800, height 1.27m, with a thermometer for correcting the barometric height before converting to altitude. The folded tripod becomes a carrying box. Geneva, Musée d'Histoire des Sciences.

161 (*right*) Two examples of 18th-century clinometers, signed 'I.R.', length 322 mm, and 'J: Kley Fecit ROTTERDAM', length 143mm. Each works by having a pointer attached to a pivoted weight. Cambridge, Whipple Museum.

162 Two types of artificial horizon: a mercury trough with glazed cover (and a turned wooden bottle for the mercury), early 19th century, by Troughton; a black glass mirror in a brass mount, with three levelling screws and a bubble level, mid-19th century, by Troughton & Simms. The artificial horizon was used on land to measure the altitude of a heavenly body with an instrument such as a sextant. The angle between the body and its reflection in a plane, level surface was measured and divided by two. Cambridge, Whipple Museum.

Surveying sextants

A sextant intended for use on land (fig. 99) can be distinguished in several ways from the nautical instrument. Most obviously, it has no shades, since it was not intended for sighting the Sun. Surveying sextants tend also to be smaller and more delicately made, and eighteenth-century examples may have fine cases; they were not built to cope with the rigours of seafaring. There might also be some means of mounting the sextant horizontally on a tripod or staff. Such instruments are more commonly of continental than of English origin.

The English, on the other hand, introduced a surveying instrument at the end of the eighteenth century, based on the principle of the sextant, but very different in appearance. This was the 'box sextant' (fig. 163), sometimes ascribed to William Jones, and described by him in his 1797 edition of Adams's *Geometrical and graphical essays*. The optical components of a sextant were miniaturized and placed within a squat cylindrical brass box some 3 in in diameter. Outside the box were the index arm and divided arc, the index mirror being moved either by adjusting the arm or by rack and pinion.

Both the sextant and the reflecting circle – applied across astronomy, navigation and surveying – demonstrate the continued congruence of our three subjects. It is also indicated by the makers involved, with the same names appearing in different fields. At the close of the eighteenth century, the whole domain still represented the élite division of the instrument-making trade, perhaps more emphatically than ever, and this remained the case for another half-century or so. In the nineteenth century, however, other branches of instrument-making came to challenge the traditional dominance of practical mathematics.

163 Box sextant by W. & S. Jones, *c.*1810, diameter 70mm. The optics are enclosed within the brass box, the index mirror and arm are moved by rack-and-pinion from the knurled knob, and the tiny silvered scale is read by vernier to 1minute of arc with the aid of the magnifying glass. The table on the lid (which also screws on to act as a handle) is used for measuring heights and distances with the sextant (see Adams, 1803, pp.265–6). Cambridge, Whipple Museum.

10

The Industrial Age

IN 1800, THE YEAR OF RAMSDEN'S DEATH, THE international situation of precision instrument-making was, from a British point of view, unambiguous. The supremacy of the London workshops was firmly established, and there seemed no prospect of their position being undermined. If the English had come to regard this as a natural state of affairs – and there is evidence that they had – their complacency was ill-founded, for the seeds of change had already been sown. The resuscitation of French mathematical instrument-making had been accomplished by Lenoir. In 1801 Lalande, who had visited the English workshops, including Ramsden's, in 1788, could write (in the language of the revolutionary era) that 'Le citoyen Lenoir a fait voir à Paris, dans l'exposition publique de l'an IX, que l'industrie française ne le cède plus à celle des Anglais' (Daumas, 1953, p.365). In the following year, Georg Reichenbach (1772–1826), who had been on a similar visit to London in 1791, founded a workshop with Joseph Liebherr (1767–1840), and so inaugurated the famous tradition of the 'Munich School'.

164 Douglas reflecting protractor by Cary, mid-19th century, length 159mm. The design was patented by Sir Howard Douglas in 1811, and the instrument was used on a plane table. The motions of the sighting arm, with its 'horizon' glass, and of the index arm are linked, so that while the former moves over half the angle to be measured, as with a normal sextant, the latter describes the true angle. Lines of sight may thus be marked directly on the paper, or may be measured on the scale. There is also a linear scale of 1 mile, taking 1 yard to be the smallest transversal division.

The first overt test of strength came in 1851, with the Great Exhibition in the Crystal Palace in London. The Exhibition had an important competitive element, exhibits being divided into thirty subject classes with an international jury appointed to each class. A jury could award a Prize Medal, or a lesser Honourable Mention, but could only recommend the highest recognition, the Council Medal, whose award required the approval of the Council of Chairmen of the individual juries. Scientific instruments were placed in Class X, whose jury was chaired by Sir David Brewster (1781–1868).

The English makers were complacent in their preparation for the Exhibition, and James Glaisher (1809–1903) recalled that when it opened, 'The inadequate representation of British work soon became glaringly apparent' (Glaisher, 1852, p.401). The leading English maker was William Simms (1793–1860), who had entered into partnership with Troughton in 1826, and now headed the firm of Troughton and Simms. Simms had made no contribution, and so 'was urged and requested, late as it was, to retrieve our credit by exhibiting' (*ibid*).

Simms was able to put together a collection which, while it represented some new work, also included navigation instruments of well-established designs by Troughton, and the astronomical instrument known as the 'Westbury Circle' (fig. 182) that had been made by Troughton in 1806. The jury recommended the award of a Council Medal, but were clearly embarrassed when the Council of Chairmen refused their approval. Simms was left with the Prize Medal only, and the jury felt obliged to record in their report that 'Mr Simms will not receive that kind of Medal to which the Jury considers him fully entitled' (Exhibition, 1852, p.269).

The embarrassment of having persuaded Simms to exhibit, only to see him snubbed, was compounded on the British side, when the only Council Medals in the traditional mathematical sciences went to French, German and even American makers. Gustave Froment (1815–65) of Paris was successful with his theodolites and a divided scale, Georg Merz (1793–1876) of Munich with an equatorial instrument, and William Cranch Bond (1797–1859) of Boston for a method of recording transit observations by a pen, operated by an electro-magnet, marking paper on a rotating drum. Traugott Leberecht Ertel (1778–1858), Reichenbach's successor in Munich, also had the jury's recommendation refused by the Council and was awarded (as 'Ertel & Sohn') a Prize Medal. In navigation, Prize Medals went to two Americans, Ericsson and St John; in surveying to the American William Austin Burt (1792–1858) for his solar compass, to A. Beaulieu of Brussels and to Friedrich Wilhelm Breithaupt (1780–1855) of Kassel.

English makers, by far the largest group of contributors, naturally carried off the majority of the medals; they had been successful in some areas of science, such as microscopy, but their performance had been disappointing in the traditional mathematical sciences. This was in spite of the fact that the German instruments, while deserving the jury's 'highest praise', '[did] not fully represent German art' (Exhibition, 1852, p.248), or as Glaisher put it, 'Germany by no means put forth her strength' (Lectures, 1852, p.348). Lyon

165 Circumferentor by E.T. Newton, 19th century, diameter 216mm. The double slit-and-window sights can be folded down for stowing. With the instrument in the vertical plane, the pivoted index beneath the needle acts as a clinometer. A single circular bubble level replaces the more usual arrangement of two long bubbles at right angles. Made in the mining district of Cornwall, this instrument may properly be called a 'miner's dial'.

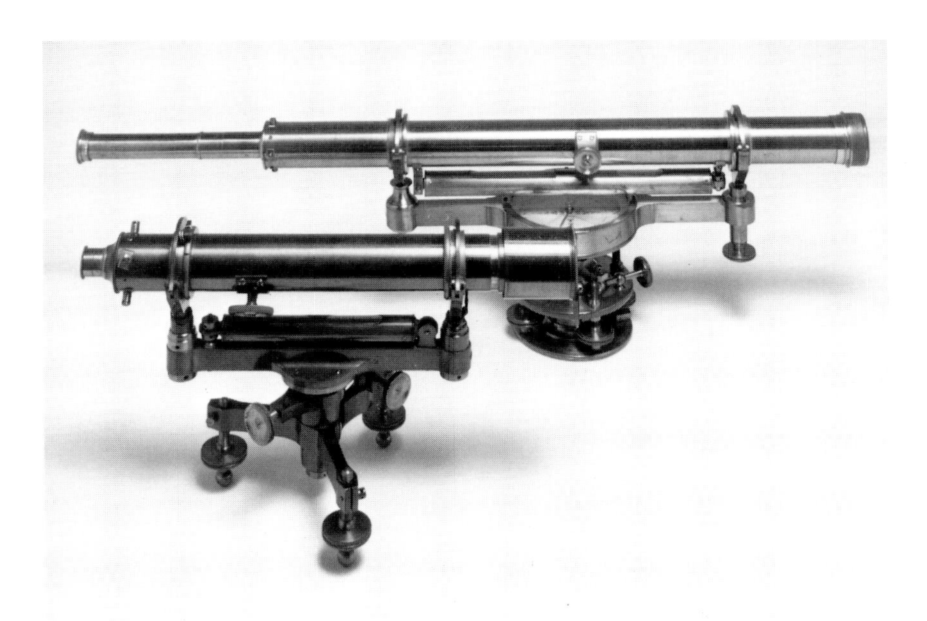

166a Two Y levels by Troughton & Simms: in front, *c.*1900, length 357mm; behind *c.*1860, length 709mm. Note that the later instrument is shorter, has no compass, and has a tribrach base with three levelling screws instead of four screws between parallel plates. Cambridge, Whipple Museum.

166b Combined compass and clinometer, *c.*1900, 76 by 76mm, with sights and bubble levels. This instrument is unsigned, but a J. H. Steward catalogue of 1928 describes the design as a 'geological' instrument, made at the suggestion of the Professor of Mines. Cambridge, Whipple Museum.

167 (*below*) Two clinometers by L. Casella: in front, *c.*1900, length 119mm (an 'Abney level'); behind *c.*1920, length 144mm (gravity clinometer). The former displays the target and by reflection, the bubble, which is adjusted level; the semicircle then gives the inclination. Cambridge, Whipple Museum.

Playfair (1818–98) summed up British consternation, when he recalled that 'Our manufacturers were justly astonished at seeing most of the foreign countries rapidly approaching and sometimes excelling us in manufactures, our own by heredity and traditional right' (Lectures, 1852, p.194). For the remainder of the century, industrial and commercial developments dominated the trade, competition was fierce, and no-one could afford to be complacent.

England

Over the first few decades of the nineteenth century, English makers of mathematical instruments lost their standing in the national scientific community. The leading workshops were Dollond and especially Troughton and Simms, and the leading figures after Troughton's death were George Dollond (1774–1852) and William Simms. Both became Fellows of the Royal Society, as did Thomas Jones (1775–1852), who had worked under Ramsden and had been on good terms with Troughton. These were the last London instrument-makers to be elected to fellowship, and papers on instrumentation gradually disappeared from the *Philosophical transactions*. Indeed the contribution of the makers to scientific literature as a whole declined.

The Astronomical Society, founded in 1820 and granted royal status in 1823, at first provided a more open alternative, with Troughton, Dollond, Jones and Simms each elected to the governing Council during the early decades, and papers from Troughton, Dollond and Simms appearing in the journals. The subsequent contribution of the makers was thin, and the tradition of including an instrument-maker on the Society's Council was continued during the third quarter of the century, but then lapsed.

Neither the Royal Society nor the Royal Astronomical Society awarded prizes or other honours to instrument-makers. The Society for the Encouragemenet of Arts, Manufactures and Commerce (the Society of Arts), founded in 1754, offered an alternative outlet for an ambitious maker, and one who took particular advantage of this was James Allan (*fl.*1790–1820). Allan was awarded a Gold Medal for a dividing engine in 1810, a Silver Medal for a reflecting circle in 1811, a Gold Medal in 1816 for a theodolite and a Silver Medal the same year for a screw-cutting engine. A final Silver Medal was awarded posthumously in 1821 for improvements in dividing. This kind of recognition was useful to aspiring makers, but was no substitute for close relationships with the scientists themselves.

Other notable makers in the first half of the century include John Stancliffe (*fl.*1770–1810), Charles Schmalcalder (*fl.*1806–38) and Matthew Berge, none of whom were survived by enduring workshops. William Cary (1759–1825) and Henry Hughes (*fl.*1836–79), on the other hand, established businesses that flourished well into the present century. Survival depended more on commercial acumen than on scientific originality, with the most successful concerns adopting industrial procedures, such as establishing moderately-sized factories, with individual workshops catering for specialist processes – casting, grinding, turning, polishing, dividing, finishing, and so on. In addition, extensive catalogues offered a large range of instruments, many of which would not be made in the factory indicated by the signature. Negretti & Zambra, for example, noted for the manufacture of meteorological instruments, published catalogues listing over 4, 000 items, among them a full range of instruments for astronomy, navigation and surveying.

As well as Hughes and Son, other large manufacturers towards the end on the century were Spencer, Browning and Co., Elliott Brothers and Heath and Co. All were particularly associated with navigation instruments. In surveying, William Ford Stanley (1829–1909) had founded one of the most important companies. Their work was very fine, with high standards maintained in the manufacture of large numbers of instruments, but there was little of

168 (*above*) Waywiser by Elliott Brothers, 19th century, height 1.18m, with an iron frame and wheel and mahogany handle, displaying yards and furlongs.

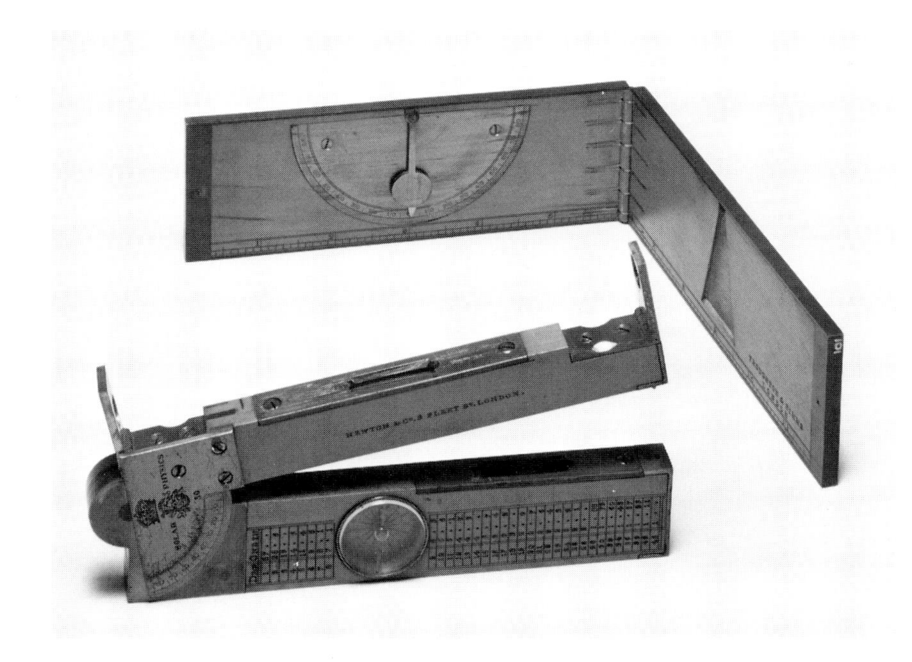

169 (*above left*) Two boxwood 'rule' clinometers: in front by Newton & Co., *c.*1900, length 305mm (1 ft), with altitude quadrant, sights, bubble levels and compass; behind by Troughton & Simms, *c.*1870, length 305mm (1 ft), with plum-bob clinometer. Cambridge, Whipple Museum.

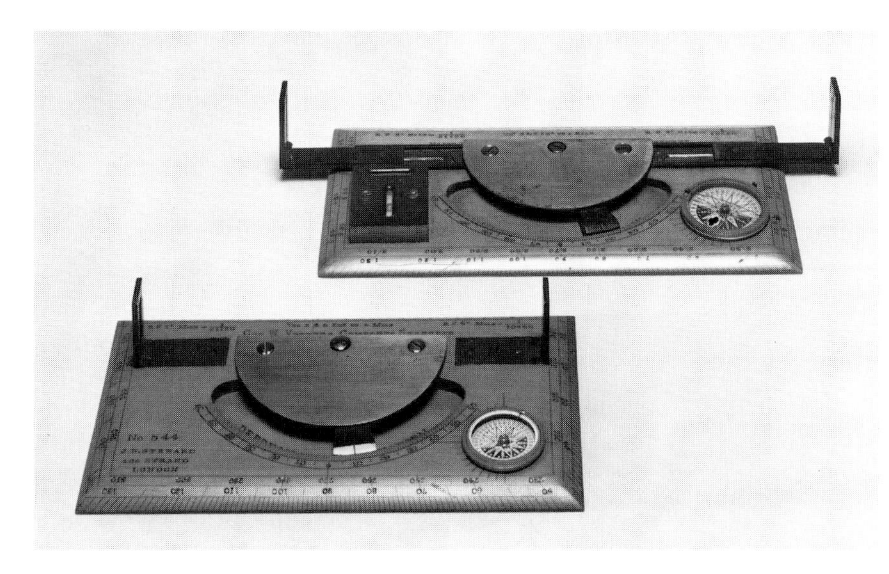

170 (*left*) Two examples of Verner's sketcher. Both are by J.H. Steward, the one in front, *c.*1925, length 141mm, is stamped 'COL. W. VERNER'S COMPLETE SKETCHER', but the one behind, *c.*1929, length 153mm, is credited to 'MAJOR W. VERNER'. Cambridge, Whipple Museum

the old excitement and originality. It was generally a period of consolidation rather than invention; improvements originated by the makers were technical rather than inspirational, such as new alloys and new frames for sextants, and rationalized and standardized designs for theodolites and levels. If scientific prestige was still indicated by the commissions of the national observatory, traditional precision work was exclusively placed with Troughton and Simms.

Germany

Breithaupt of Kassel was the only German workshop that survived and prospered into the nineteenth century; as F. W. Breithaupt & Son they were awarded a Prize Medal at the Great Exhibition. In the early years of the century, however, a revival of German instrument-making was begun in Munich, and would establish a tradition of unprecedented importance since the days of the best makers of Nuremberg and Augsburg. This revival has been linked to political developments in contemporary Germany – the encouragement of the Bavarian government, the land surveying activity occasioned by the Congress of Vienna in 1815 and, particularly, the founding of the German Empire in 1871, after which enlarged markets allowed a tradition of ingenuity and craftsmanship to be consolidated into a commercial business with its centre in Berlin.

The names of the principal members of the 'Munich School' – Reichenbach, Liebherr, Utzschneider, Fraunhofer, Ertel, Merz and Steinheil – appear on instruments in a confusing variety of combinations. The sequence of partnerships has been explained by A. Brachner (1985). In 1802 Reichenbach and Leibherr founded a 'Mathematical-mechanical Institute' in Munich, for the manufacture of instruments for astronomy and surveying ('Reichenbach & Liebherr'). They were joined by Joseph von Utzschneider (1763–1840) in 1804 ('Reichenbach & Utzschneider & Liebherr'), and by Joseph von Fraunhofer (1787–1826) in 1806. In 1809 a second 'Optical-mechanical Institut' was founded in Benediktbeuern by Utzschneider, Reichenbach and Fraunhofer, which became 'Utzschneider & Fraunhofer' after Reichenbach's withdrawal in 1814. In the same year, Liebherr retired and Utzschneider withdrew

171 Compass card by G. Hulst van Keulen of Amsterdam, early 19th century, with an adjustment for variation. Amsterdam, Nederlands Scheepvaart Museum.

172 (*below*) Sextant by G. Hulst van Kuelen, *c.*1800. This is an unusual type of frame and Dutch sextants in general are uncommon. Amsterdam, Nederlands Scheepvaart Museum.

173 (*right*) Dividing engine by Reichenbach, 1804. Munich, Deutsches Museum.

174 (*opposite*) Theodolite by Ertel & Sohn, *c.*1840. The double horizontal circle allows repetition of observations. The vertical circle is balanced by a counterpoise. The tangent screws acting against enclosed springs are clearly seen – two for the horizontal circle, one for the vertical. London, Science Museum.

München

175 (*above*) Repeating theodolite by Reichenbach, 1810. The two telescopes and circles allow repetitions of a measurement, the total angle being divided by their number; this reduces errors of targetting, scale reading, scale division and eccentricity. Munich, Deutsches Museum.

from the original Munich Institute, leaving it with Reichenbach and Traugott Leberecht Ertel ('Reichenbach & Ertel'). Ertel took over the management when Reichenbach retired in 1820, and was joined by his son Georg (1813–63) in 1834 ('Ertel & Sohn').

After Utzschneider's death in 1838, the Benediktbeuern workshop was owned by Georg Merz and Joseph Mahler (1795–1845), who had effectively managed it since Fraunhofer's death in 1826. Merz was the sole owner from 1845, and was succeeded by his son Sigmund in 1867. The workshop of Carl August Steinheil (1801–70) was established independently in 1855; Steinheil worked from the beginning with his son Adolph (1832–92), and with a second son Eduard (1830–78) from 1862. The business passed to Adolph's son Rudolph (1865–1930) in 1892. While many other people were involved, these were the best-known branches of the Munich workshops.

Johann Georg Repsold (1771–1830) began his workshop in Hamburg around 1799. As with the Munich School, his work was mainly concerned with astronomy and surveying, and he concentrated on a relatively small number of very fine instruments. His sons Georg (1804–85) and Adolf (1806–71) succeeded to the business in 1830 ('A. & G. Repsold'), and expanded the trade, though still with an emphasis on astronomy and surveying. A change of name to 'A. Repsold & Söhne' in 1867 marked the retirement of Georg and the partnership of two of Adolph's sons.

Two Berlin workshops active in astronomy and surveying in the early part of the century are worth noting – those of Carl Philipp Heinrich Pistor (1778–1847), founded in about 1813, and of Johann August Daniel Oertling (1803–66) in about 1826. Trading as 'Pistor

& Martins', the former also made navigation instruments, which had not been typical of the German makers of the nineteenth century. The market for navigation instruments was greatly expanded at the beginning of the Empire in 1871 and the establishment of the Imperial Navy. In that same year, a workshop which took particular advantage of this opportunity was founded by Carl Johann Wilhelm Bamberg (1847–92). Maintaining a large workshop and an extensive range of commercial instruments, Bamberg represented the mathematical side of the Berlin trade, which dominated German instrument-making by the end of the century.

178 (*left*) Dividing engine by Oertling, 1840 Munich, Deutsches Museum.

179 (*right*) Prismatic quintant by Pistor & Martins, mid-19th century, with a prism replacing the normal horizon glass. The arc is divided up to 250°. Amsterdam, Nederlands Scheepvaart Museum.

France

Three French mathematical instrument-makers – Lenoir, Fortin and Jecker – were prominent in the early nineteenth century. From around the first decade, Lenoir's workshop was in the charge of his son Paul Etienne (1776–1827), though the elder Lenoir remained active, and invented a form of dividing engine in 1825. Nicolas Fortin (1750–1831) is associated more with a range of physical and chemical apparatus than with mathematical instruments, but late in his career, he became successfully involved with astronomy and surveying. The importance of François-Antoine Jecker (*fl.*1790–1820) lies in his establishment, in about 1800, of a large workshop for the efficient production of numbers of low-cost instruments, among them sextants and reflecting circles. He employed some forty men, and had benefited from several years' experience in the only slightly larger workshop of Ramsden.

Noël-Jean Lerebours (1761–1840) made some mathematical instruments, but was better known as an optician. Henri Prudence Gambey (1787–1847) was probably the most important mathematical instrument-maker in the second quarter of the century, making both large-scale astronomical instruments and fine instruments for surveying and navigation, such as theodolites, sextants and reflecting circles.

176 (*opposite, left*) Universal instrument of Utzschneider & Fraunhofer, *c.* 1825, diameter 245mm. In this configuration the instrument is an altazimuth theodolite with short vertical arcs and a horizontal circle divided to 10minutes of arc and read by vernier to 10 seconds. The horizontal circle has four index arms and four reading microscopes; the vertical has two. The usused pieces are for converting the instrument into a vertical circle (fig. 177). Cambridge, Whipple Museum.

177 (*opposite, right*) The universal instrument in fig. 176 remounted as a vertical circle. The telescope used in this configuration is missing. Cambridge, Whipple Museum.

Noël-Marie Paymel Lerebours (1794–1855), son of Noël-Jean, and Marc Secretan (d.1867) formed a partnership in 1847. Secretan's interests directed production towards mathematics, and in 1855 he took over the workshop, which was now able to undertake large commissions from the Paris Observatory. It remained in the Secretan family into the twentieth century and established a large catalogue of instruments in astronomy, navigation and surveying – equatorials, portable transit instruments, theodolites, reflecting circles, and so forth.

Other important workshops specializing in mathematical instruments at the end of the century included Balbreck Aîné & Fils (founded in 1854, and making surveying instruments), Bellieni (1812, surveying), Brosset Frères (1855, surveying), Gautier (1876, large astronomical instruments, and geodetic instruments), Mirvault (founded by Trochain in 1854, and making instruments for surveying and navigation), Morin & Genesse (founded by H. Morin in 1880, the partnership in 1886, specializing in surveying and geodesy), Ponthus & Therrode (who took over the workshop of Berthélemy in 1895, and had a wide range of surveying and navigation instruments).

America

While most of the instruments for optics and natural philosophy used in America in the eighteenth century were manufactured in London, the work of S. A. Bedini, has uncovered many makers of mathematical instruments active in America. Some, no doubt, were mostly concerned with retailing, and among those who kept shops, many more probably held mixed stocks of locally manufactured and imported instruments. There was, however, a growing instrument-making capability in America, which in the nineteenth century developed into a manufacturing industry.

180 American graphometer, *c.*1800?. A simple instrument in wood with a trough compass and brass alidade, which fits round the edge when not in use. The semicircle is divided to 1°. Cambridge, Whipple Museum.

Some well-known early makers were Thomas Godfrey (1704–49) of Philadelphia, who devised a form of reflecting quadrant, Anthony Lamb (1703–84), a convicted felon transported from England and an early maker of Godfrey's quadrant, and Thomas Greenough (1710–85) of Boston, one of the most prolific makers. One business that extended into the nineteenth century was established by William Williams (*c.*1737–1792) in Boston in 1770. He was succeeded by Samuel Thaxter (1769–1842), who had married Williams's niece, and as 'Samuel Thaxter & Son' the business continued in Boston throughout the nineteenth century.

One of Williams's advertisements in 1770 begins by listing 'A large Assortment of Hadley's and Davis's Quadrants, hanging and standing Compasses, in Brass and Wood. Gauging and Surveying Instruments . . .'; in 1793 Thaxter advertised 'A very neat Sextant, and a large assortment of Hadley's Quadrants, Davis' d[itt]o. . . . Gauging and surveying Instruments, pocket Compasses, binnicle aud [sic] cabin [i.e. hanging] Compasses . . .' (Bedini, 1964, pp. 95, 99). The emphasis throughout the American trade was very firmly on navigation and surveying, and by far the most common instru-

181 American transit by W. & L. E. Gurley, late 19th century, height 135mm, with full vertical and enclosed horizontal circles, in addition to the large compass characteristic of this type of instrument.

ment was the circumferentor, in brass or wood. A few makers were able to produce fine astronomical instruments, notably two clock-makers of Pennsylvania, David Rittenhouse (1732–96), and Andrew Ellicott (1754–1820). Rittenhouse's brother Benjamin (1740–*c.*1820) also made clocks and surveying instruments.

The nineteenth century saw the development of this individual workshop tradition into the kind of industrial and commercial concerns we have already met in Europe. An important stimulus was the foundation of the United States Coast Survey, which began work in 1816, and represented the earliest major government sponsorship for science in America. A general book cannot do justice to the American makers, but we will mention a few.

The workshop established by Benjamin Pike (1777–1863) in New York in 1804, became Pike & Son in 1831, when his son Benjamin (1809–64) became a partner, and 'Pike & Sons' when joined by Daniel (1815–93) in 1841. In 1843 Benjamin, Jr. left to set up on his own, while the older firm became 'Pike & Sons' once more when Gardiner (1824–93) joined in 1850. Instruments signed 'B. Pike's Son' (as distinct from B. Pike, Jr.) were made after the death of Pike, Sr., and the withdrawal of Gardiner. In 1848 Benjamin Pike, Jr. published a large catalogue (second edition 1856) with a very extensive range of instruments. He included valuable descriptive and explanatory accounts of the instruments, while reminding readers of the preface that 'He wishes it borne in mind, that he is not a man of letters, but a mechanic, – a practical workman' (Pike, 1848, p.v). Most surviving Pike instruments are for surveying.

Two New York workshops involved in the marine chronometer trade also supplied navigational instruments: John Bliss (1795–1857), which survived its founder as John Bliss & Co., and E. & G. W. Blunt, a partnership between brothers Edmund (1799–1866) and George William (1802–78). William J. Young (1800–66) came to America from Scotland and was apprenticed in 1813 in Philadelphia to another immigrant instrument-maker, Thomas Whitney (d.1823) from London. Young established his own workshop around the time of Whitney's death, and built the first dividing engine in America. As Young & Sons, the firm traded into the twentieth century, and specialized in surveying instruments.

One of the largest firms, to judge from the numbers of surviving instruments, was W. & L. E. Gurley, of Troy, New York. It derived from the workshop founded by Jonas H. Phelps (1809–65), who was joined by William Gurley (1821–87) in 1845, and by a brother Lewis E. Gurley (1826–97) in 1851. Phelps retired in 1852. Surveying instruments again predominate, and the company has continued in business to the present day.

William Cranch Bond built the first American marine chronometer in 1812, working in the chronometer shop of his father William in Boston. He was also interested in astronomy, and it was his technique for registering transit observations by an electrically operated pen that won a Council Medal at the Great Exhibition.

The 1851 exhibition indicated that the international distribution of expertise and originality in mathematical instrument-making had radically changed, and that a period of more balanced competition was bound to ensue. It also displayed new techniques that would alter fundamentally the position of the divided circle tradition in science: photography, which in time would have a profound influence on the practices of both astronomy and surveying, and the electric telegraph, which was soon to yield a novel technique for distributing time signals. The entirely new analytical technique of spectroscopy was displayed at the next international exhibition in London, in 1862. After thousands of years, man had discovered that the light from the stars carried with it information about the nature, as well as the position, of its origin. This insight powerfully influenced the research goals of astronomy, which had always motivated developments in the technology of the divided circle.

11

Astronomical Circles

THE NINETEENTH CENTURY SAW BOTH the zenith and the decline of the divided circle in astronomy. At the beginning of the period fundamental precision astronomy was considered the primary duty of an observatory, and the new circular instruments were revising once again the accepted standards of accuracy. The status of their makers had never been higher. They were not handed commissions to execute; rather they helped to decide on appropriate instrumentation, and those at the top of the profession, such as Ramsden or Troughton, were able to exert a very considerable influence on the development of astronomy.

This was a role the leading makers came to expect. In 1784 Hornsby began to refer ironically in his correspondence to 'le Sieur Ramsden'. When the amateur astronomer Francis Wollaston wished to order a transit circle in 1788, he applied to Ramsden, only to find that '. . . the multiplicity of his engagements and the fertility of his own imagination . . . rendered him disinclined to listen to a scheme for one on another plan'. 'The same was the case with Mr Troughton', as Wollaston discovered, but he remained convinced that 'Observers know best what it is they want; and an instrument-maker who will condescend to listen to them, is a treasure' (Dewhirst, 1986, p.154). The very first two papers published in the *Memoirs of the Astronomical Society* in 1822 were by Edward Troughton and George Dollond. In 1829 William Pearson, FRS, prefaced his definitive account of astronomical instruments with a printed dedication to Troughton, 'as a testimony of my great esteem for your extraordinary talents' (Pearson, 1829).

Such was the enthusiasm for precision astronomy, and the prestige attached to it, that the fashion for founding observatories was greater than ever. The functions of official observatories in particular – those founded by national or local governments or by universities – were to a large degree symbolic. An impressive observatory building indicated an enlightened interest in the highest form of science. Inside were placed the latest models of precision instruments, and the staff settled down to extended programmes of meridian observations. There was generally no clear theoretical aim to the exercise, and the observations themselves often remained unpublished, or if published remained unused.

182 Troughton's 'Westbury circle', made for John Pond in 1806. (Pearson, 1829)

Not that this was the only kind of astronomy there was. The last great period of observational astronomy – the quest to know what the heavenly bodies were, rather than where they were – had been in the seventeenth century, with the first application of the refractor to astronomy. At the end of the eighteenth century, William Herschel (1738–1822) had pioneered the serious application of the reflector to discovering the heavens, and had built enormous telescopes for penetrating far into space. Herschel, of course, had been an amateur, standing outside the professional community and its understanding of the nature of astronomy, and his field remained the province of enthusiastic amateurs for much of the nineteenth century. Foundation and management committees, boards of trustees and visitors accepted the consensus view and authorized the solid conventions of neo-classical architecture and precision instruments.

Altazimuth circles

Two types of astronomical circle were constructed at the end of the eighteenth century. The altazimuth circle was generally a portable instrument, with the notable exception of Ramsden's Palermo circle. The transit circle (or 'meridian circle'), on the other hand, was a meridian instrument, intended for measuring declinations by the divided circle at the same time as right ascension was noted by the clock. The Dublin circle was of this type. Another important early example, though of only 2 ft diameter, was the instrument eventually ordered by Wollaston from William Cary, who was less famous and more amenable than Ramsden or Troughton.

183 Troughton's 'South Kilworth circle' of 1821. (Pearson, 1829)

Altazimuth circles continued to be built on a relatively modest scale. A well-known instrument, called the 'Westbury circle' (fig. 182), was made in 1806 by Troughton for the private observatory of John Pond (1767–1836), later Astronomer Royal. It had the now-familiar arrangement of azimuth circle (2 ft diameter) carried by radial tubes, and double altitude circle (30 in diameter), each divided to five minutes and read by two micrometer microscopes. The horizontal axis and vertical circle were carried by two tapering pillars, the whole set in a stone pier, with an elaborate arrangement for establishing a true vertical axis. This may have been found necessary at a late stage in the design, since the instrument was originally intended as an equatorial.

A slightly larger altazimuth circle (fig. 183) was ordered from Troughton by the Royal Academy of St Petersburg, but cancelled with the instrument almost finished in 1821. Bought by Pearson, to become the 'South Kilworth circle', it had 3 ft circles, the azimuth read by three micrometer microscopes, the vertical by four. The vertical circle was double and raised on two columns as before, but the vertical axis was now set more conventionally, by a tripod base with three levelling screws.

From instruments such as these evolved the portable altazimuth circle, of lesser dimensions and generally used in a smaller private observatory. There was never a standard pattern, but sometimes a single column and single vertical circle were used, and verniers com-

184 (*left*) Portable altazimuth by Dollond, *c.*1790, diameter (horizontal circle) 143mm. The division of both circles is to 30 minutes of arc, read by vernier to 1minute. Cambridge, Whipple Museum.

185 (*right*) Repeating circle by Thomas Jones, *c.*1835, diameter (main circle) 295mm. A typical repeating circle, with a tribrach base and azimuth circle, and a fork mount for the double circle with two telescopes and counter-weight. Edinburgh, Royal Museum of Scotland.

monly replaced micrometer microscopes. (Reading microscopes were for distinguishing the scales, not for measurement.) Two telescopes might be fitted, to make the instrument a repeating circle (fig. 185).

A portable altazimuth by Troughton (similar to fig. 230) was described by F. W. Simms in his *Treatise on the principal mathematical instruments* of 1838, which became the standard manual of the mid-century period and was published in a number of editions. A solid circular base plate with a divided azimuth scale was adjustable by three levelling screws and an azimuth tangent screw. Above, a close-fitting plate on a conical bearing carried two slightly tapering pillars; attached to the pillars (to the horizontal plate in fig. 230) were two micrometer microscopes for reading the horizontal scale, and the pillars supported Y bearings for the double-cone horizontal axis of the telescope. The altitude circle was in two elements, connected by bars, and was read by two micrometer microscopes mounted on one of the pillars. A striding level surmounted the horizontal axis; a hanging level linked the two upper microscopes. In 1838 the cost of

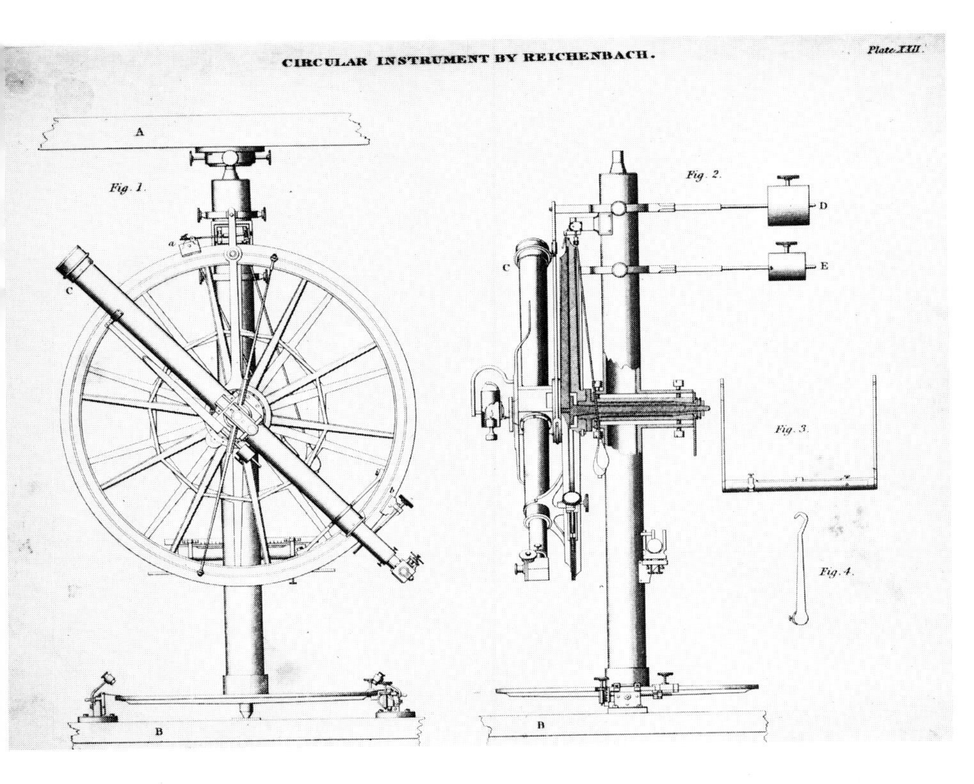

186 (*left*) One of Reichenbach's altazimuth instruments made for Naples Observatory in 1814 (Pearson, 1819)

187 (*right*) Vertical or altitude circle by Reichenbach, with a small azimuth circle. Munich, Deutsches Museum.

such an instrument from Troughton and Simms with 18 in circles was 200 guineas (£210).

In Germany, Reichenbach made portable altazimuth circles, generally on stout pillar-and-tripod or tripod stands. They were equally useful for small observatories, for geodetic surveying or for field astronomy. On a larger scale, a pair of instruments to a very different design (fig. 186) were made for the Naples Observatory in 1814. The vertical axis was a hollow brass cylinder, located in pivots on substantial metal plates top and bottom, with the plates joined by two columns clear of the rotating circle. The mount for the horizontal axis was let into the vertical cylinder. It carried two concentric circles, the outer of 1 m diameter and divided to five minutes, the inner linked to the motion of the telescope and having four verniers reading to two seconds. The azimuth circle at the base of the vertical cylinder was divided to five minutes and read by two verniers to four seconds. Counterpoises were used to distribute the weights on the telescope side equally across the vertical axis.

Reichenbach's circle may well have seemed strange to an Englishman. Pearson significantly began his critique by establishing a framework of national traditions:

> One of the first objects of remark that occurs to an English astronomer . . . is, that the readings of the circles are by means of verniers, now that the use of micrometrical microscopes is become general in England, whenever instruments are made on a scale large enough to admit of accurate subdivisions to spaces of 5′. (Pearson, 1829, p.491)

Reichenbach had served notice that he was not bound by standard designs, taken over from the dominant English school, but was prepared to experiment.

Mural circles

Nowhere did the difference between the English and German schools of instrument-making, represented early in the century by Troughton and Reichenbach, have more effect in astronomy than on the principal meridian instruments of observatories. While Reichenbach made transit circles, Troughton introduced an instrument of his own design, intended only for the measurement of declinations (or zenith distances). This was the mural circle, which had, of course, to be accompanied by a transit instrument for measuring right ascensions. This type of instrument, then, was the nineteenth-century equivalent of the mural quadrant.

The paradigm mural circle (fig. 188), though not the first to be completed by Troughton, was that supplied to the Greenwich Observatory in 1812. Before then, declination measurements were still being taken with a mural quadrant built by Bird, and essentially designed by Graham, and in 1806 Maskelyne had pointed out to the Royal Society that it was high time the national observatory had a large circle:

> . . . in consequence of the great improvements that have been made in the last twenty years, principally by British artists, in the construction of astronomical instruments, most of the Observatories in foreign countries are now furnished with divided circles for observing the distances of celestial objects from the Zenith . . . (Howse, 1975, p.28).

188 Mural circle by Troughton, 1812, diameter 6 ft (1.8m), for the Royal Observatory, Greenwich. All six original micrometer microscopes are missing.
Greenwich, National Maritime Museum.

Troughton made a model of the proposed mural circle in the same year, and it may be that the original context for this instrument – a replacement for the mural quadrant, for measuring zenith distances (or declinations) – influenced the design. The idea was to apply the proven advantages of the circle in upgrading the quadrant, which made the 'mural circle' a natural outcome. Maskelyne, who died in 1811, had hoped that the circle might be used for transits also, but it proved unsuitable. The necessary adjustments were more easily made with a transit instrument, and a new one (fig. 190) was ordered from Troughton in 1813. Troughton was now convinced – in part, it might be suggested, by the progression of events at Greenwich – that the two functions should be separated in two instruments, so as to maximize the advantages of stability and adjustment appropriate to each case. He inscribed on his transit instrument, 'To the President and Council of the Royal Society this and the Mural Circle, being his greatest and best works, are dedicated by the maker' (Howse, 1975, p.38); the pair of principal instruments at the national observatory had established a precedent that would be difficult to deny.

The Greenwich circle, with optics by Dollond, was 6 ft in diameter. A 4 ft-thick meridian wall carried the adjustable mount for the horizontal axis, and six micrometer microscopes placed at 60–degree intervals. The circle was carried by sixteen conical radii, and comprised a flat ring parallel to the wall and an attached outer ring at right angles. This latter ring carried the scale divided to five minutes, and the microscopes lay parallel to the wall. The circle rotated with the telescope; it could be read to seconds by the wires of the micrometers, and to tenths of a second by the micrometer heads. The zero was taken, not from a plumb-line, but from the pole, located by observations of circumpolar stars.

Pearson was close to Troughton, and he treated this instrument with particular reverence; at times his language is almost mystical. Describing the genesis of the design, he wrote that 'The idea of a circle fixed to a wall, in the plane of the meridian, did not escape the artist; the steadiness, simplicity, and elegance of such a position had fixed it in his mind' (Pearson, 1928, p.473). An alternative method of establishing the zero was by observing the same star directly and by reflection in a mercury trough. 'The same laws of nature which point the plumb-line', wrote Pearson, 'form also the surface of the fluid at right angles to it. But the plumb-line is the work of man, and the reflecting surface the appointment of nature as much as the polar zero' (*ibid*).

Before the Greenwich instrument was erected, Troughton had made two smaller mural circles for observatories in Scotland: a 5 ft circle for the Garnet Hill Observatory in Glasgow, and one of 2 ft (accompanied by a transit instrument) for the private observatory of Sir Thomas Makdougall Brisbane (1773–1860). Gambey made an instrument very similar to the Greenwich example, with an accompanying transit instrument, for the Paris Observatory in 1819, but mural circles were more typically manufactured in London. They were generally, though not exclusively, installed in observatories in

190 (*opposite*) Troughton's Greenwich transit instrument, 1816 (Pearson, 1829)

Britain or in her colonies overseas. Instruments were built by Troughton and Simms for the universities of Cambridge and Edinburgh, and for the Lucknow, Madras and Trivandrum observatories in India. They were also made by Thomas Jones, who supplied the Armagh and Cape observatories, and built a companion instrument to the Greenwich circle in 1823, to provide for the simultaneous observation of stars directly and by reflection. Pearson noted '. . . the same cordiality between them, as there has subsisted between their respective makers for many years' (Pearson, 1828, p.475).

189 Transit instrument of a type designed by Latimer Clark, incorporating a reflecting prism, for domestic timekeeping, made by A. J. Frost, London, 1884, diameter (base) 162mm. Cambridge, Whipple Museum.

Transit instruments

The transit instrument was the natural complement to the mural circle, but the leading makers in England, Germany and France received orders for observatory transit instruments until the mid-century or so, from institutions with or without mural circles. The pattern was now well established, with common additions in the nineteenth century being small setting circles with bubble levels, mounted on the telescope close to the eyepiece, and counterpoises acting at both ends of the axis to reduce friction on the pivots and bearings.

The instrument was to have a longer future as a portable device,

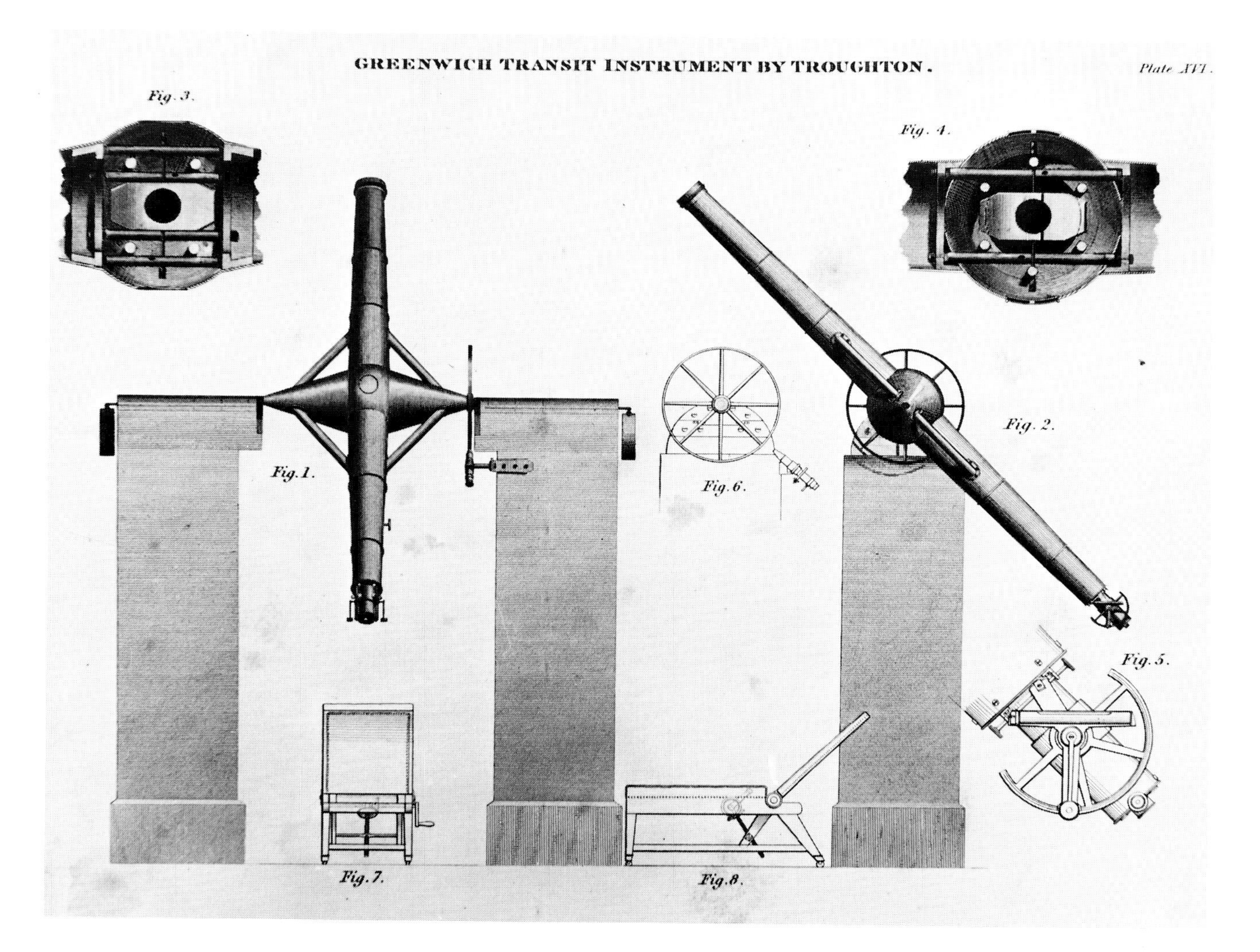

191 (*right*) 'Portable' transit instrument by Brauer, 19th century. An example of a 'broken axis' or 'axis-view' instrument, where the rays of light are diverted and the object viewed through the horizontal axis. Note the substantial cast stand, the apparatus for raising and reversing the horizontal axis, and the micrometer eyepiece. There are counter-weights for the telescope and vertical circle. London, Science Museum.

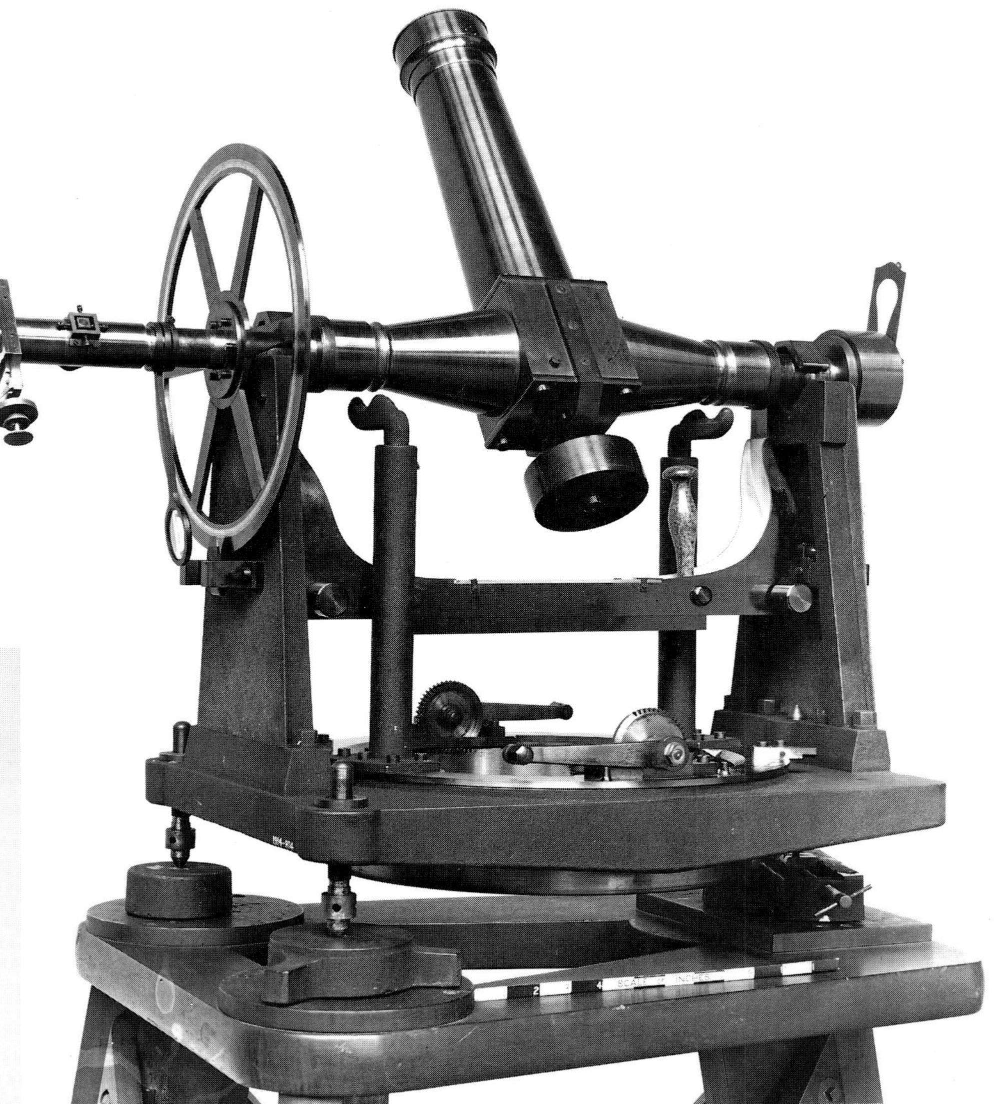

192 Portable transit instrument by Dollond, early 19th century, diameter (base) 146mm. The near side has a small setting circle and clamp, the far side a platform for a lamp, which shines along the axis and by reflection down the tube to illuminate the cross-wires. The stand is held together by knurled screws and is easily dismantled. Cambridge, Whipple Museum.

for finding the time at a temporary observing station or an observatory without a major meridian instrument, or in geodetic work for finding longitude with the help of a chronometer. Two types of portable transit instrument were fairly common throughout the nineteenth century. One (fig. 192) was based on a brass or iron ring with three levelling screws. Two A frames rose from either side of a diametric bar, each braced to the bar by a strut. The double-cone horizontal axis was reversible in Y bearings, and was surmounted by a striding level. Illumination for the cross-wires was generally introduced through the axis. At one end of the axis, and linked to its motion, was a small setting circle read by a diametric double index arm, the arm being set to the horizontal by a spirit level and then fixed by a clamping screw.

Pearson attributed the portable transit of this type to Troughton, but early examples were also made by Dollond, and a very similar type by J.G. Repsold as early as 1818. F.W. Simms described it as Troughton's model in 1838, and in his illustration the components of the stand are held together by screws with knurled heads, to be easily dismounted and packed away. It was illustrated in this form in the standard textbooks, and offered for sale in the catalogues of general makers, along with any number of other instruments. It appears, for example, in the 1848 'illustrated descriptive catalogue' of Benjamin Pike, Jr. of New York, and in the 1886 'encyclopaedic illustrated and descriptive reference catalogue' of Negretti & Zambra – in each case duly engraved with the maker's, or more probably retailer's, name.

The second type (fig. 194) had a more substantial stand, with a solid base and solid or openwork uprights and braces in a single casting. An early example by Thomas Jones was described by Pearson, and it became the more common form as the century passed. Available from almost every appropriate maker, there was a large range of sizes and considerable variety in design, from small instruments for domestic time-checking to sophisticated commissioned work for astronomical expeditions. Axis illumination was customary, small setting circles linked to the index were more common than those

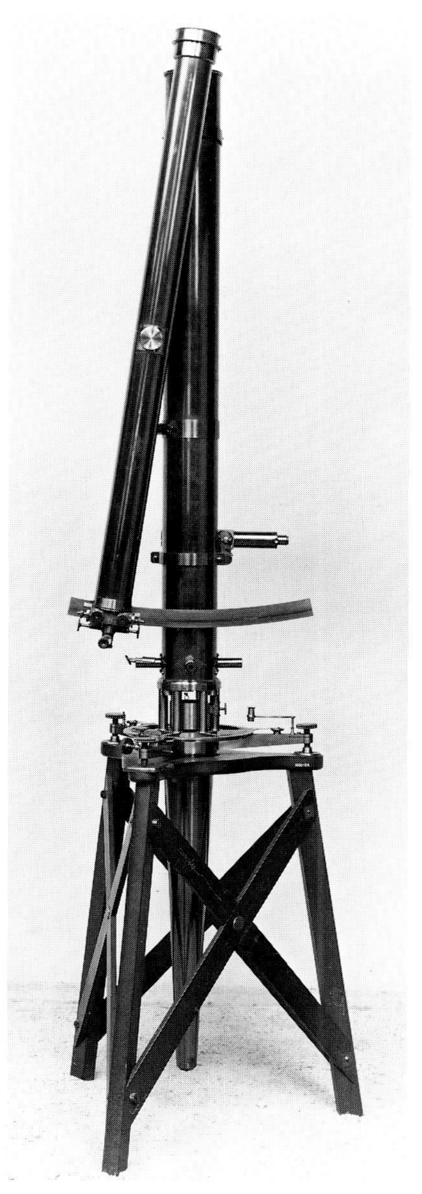

193 (*above*) Portable zenith sector by Troughton, early 19th century. The vertical arc and the conical bearing for the azimuth motion are clearly shown. London, Science Museum.

194 (*left*) Portable transit instrument by Troughton & Simms, mid-19th century, with a striding level and a lamp for axis illumination. The small vertical circle is a setting circle with tangent screw adjustment, and the bulge at the other end of the axis compensates for its weight. The stand has been made in a single casting. London, Science Museum.

195 Altazimuth instrument by Ertel & Sohn, *c.* 1840, diameter 265mm, with axis view. Both vertical circles are divided, one to 15minutes of arc, read by vernier to 1minute, the other to 10minutes and read by four verniers to 10 seconds, with the help of reading microscopes. Few portable instruments have complete provenances, but we know that this one was bought from Ertel by E. J. Cooper for his observatory in Ireland, and after his death bought by J. C. Adams for Cambridge. Cambridge, Whipple Museum.

mounted at the eye-end of the tube, striding levels were more typical than hanging, but it would be inappropriate to deal with all the variations here. As with the observatory instruments, it was more common for French or German than for English makers to expand the vertical circle and offer a portable transit circle.

Transit circles

Separate mural circle and transit instruments were expensive to build and to man, and the alternative single instrument was the transit circle, for measuring both right ascension and declination. The idea was not unknown among English makers: Cary had made one for Wollaston, and another (fig. 196) was built by Troughton in 1806 for Stephen Groombridge (1755–1832). This was a 4 ft diameter double circle, divided to five minutes and read by micrometer microscopes, and attached by conical radii to the 3 ft double-cone horizontal axis of a 5 ft telescope, the axis resting in adjustable Y bearings carried by two stone piers. This was a combination of familiar elements, arranged to produce a new instrument.

An interesting compromise (fig. 197) was provided for the Radcliffe Observatory in Oxford by Thomas Jones in 1836. This was a 6 ft circle, having much of the appearance of a mural instrument. The two elements of the double circle were linked by bars, with alternate bars connected to the centre by radial cones in the familiar manner. The mount for the central pivot was let into a substantial stone wall, which also carried the micrometer microscopes. On the other side of the centre, however, was the base of a 2 ft 6 in cone, which extended to a second pivot on a stone pier. The effect was to produce a transit circle, with many of the features of the mural circle.

The transit circle was, however, more commonly the work of the German makers, who re-equipped most of the important European observatories in the first half of the nineteenth century. A typical Reichenbach circle had a telescope with a double-cone horizontal axis, resting in adjustable Y bearings set on two stone piers. Two balance bars, carried by metal pillars set on the piers, were attached on the telescope side to the horizontal axis and on the outside carried counterpoising weights, whose positions were adjustable along the bars. These relieved the pivots and bearings of most of the instrument's weight. At one end of the horizontal axis a divided circle, of perhaps 3 or 4 ft diameter, rotated with the telescope and was fitted

196 (*left*) Transit circle made by Troughton for Groombridge in 1806. (Pearson, 1829)

197 (*right*) Circle by Thomas Jones, 1836, diameter 6 ft (1.8m), photographed at the Radcliffe Observatory. The original invoice for this instrument is preserved, addressed to the Radcliffe Trustees from Jones, 'Pupil of Ramsden'. It shows that the total cost was £914.16s.6d, of which £840 was the price of the instrument and the rest was due to delivery and mounting. Oxford, Museum of the History of Science.

with clamp and tangent screws. The index arms, with verniers, were set by a spirit level, and their positions read by microscopes. Light for the cross-wires was introduced along the horizontal axis, which was supplied with a substantial hanging level.

Such circles by Reichenbach, Liebherr, Ertel, Pistor and Martins, and Repsold were installed in numerous observatories in the first half of the century. The French also became interested in instruments of this type, and they were built by Gambey, Secretan and later by Eichens, Gautier and Mailhat. French manufacture was rather more important later in the century.

Meanwhile, the English makers came to terms with the demand for transit circles, and examples were made in the mid-century period and in its second half by Troughton and Simms, sometimes to replace earlier mural circles. One such instrument (fig. 199) was to have a special importance. This was the transit circle designed by the Astronomer Royal George Biddell Airy (1801–92) for the Greenwich Observatory and completed in 1850. The structure was built by the general engineering firm of Ransomes and May of Ipswich, while the optics, the division of the circle and other instrumental aspects were by Troughton and Simms. The telescope had an aperture of just over 8 in, and a focal length of 11 ft 7 in. The double-cone horizontal axis was 6 ft in length, resting on piers appropriately salvaged

198 (*above*) Meridian circle at the Paris Observatory, as illustrated in Chambers, 1890.

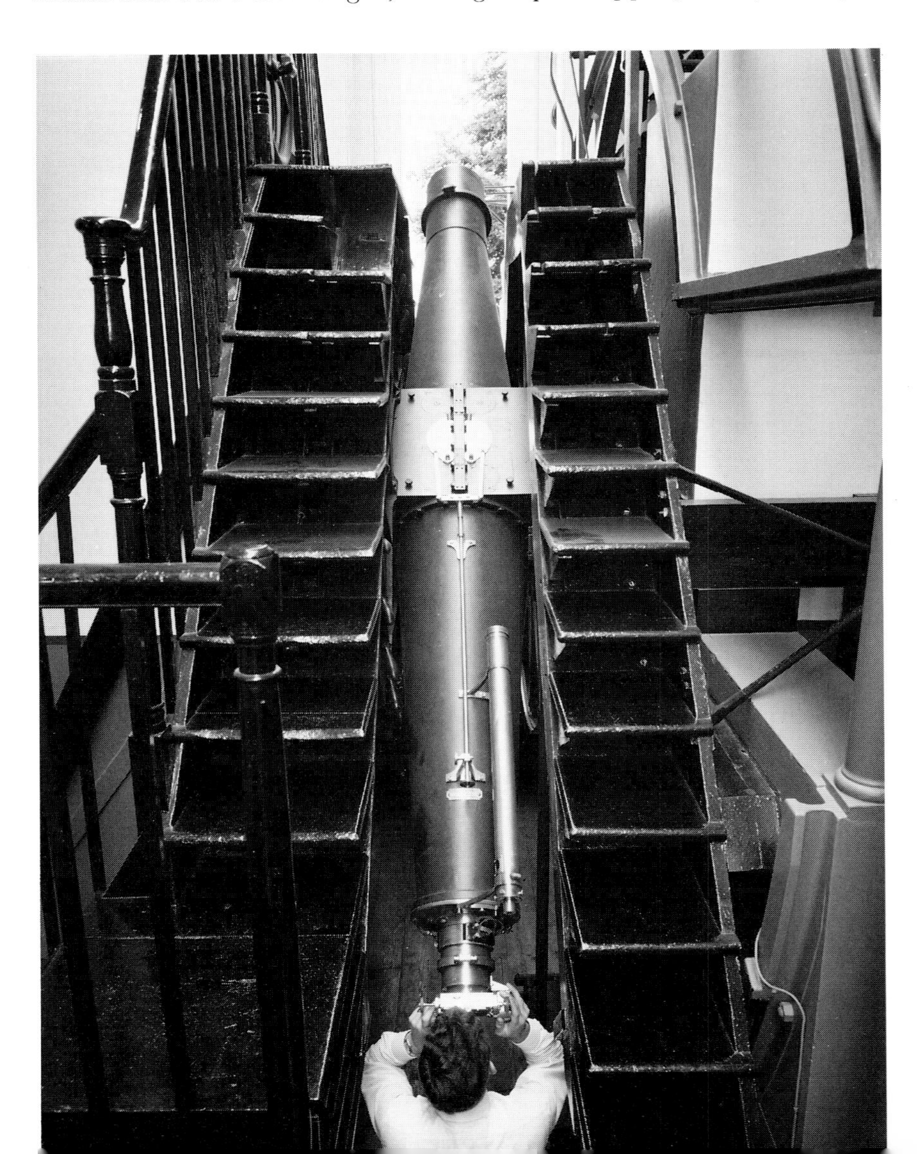

199 (*right*) The transit circle designed by Airy and installed at Greenwich in 1850, which came to mark the prime meridian of the world. Mounting by Ransomes & May, optics and division by Troughton & Simms, diameter 6 ft (1.8m). Greenwich, National Maritime Museum.

from two redundant instruments, the Troughton and Jones mural circles. The circle was 6 ft in diameter, divided on a silver scale to five minutes and read to 0.06 seconds by six micrometer microscopes, whose optical axes pierced the west pier. Counterpoises were used to reduce substantially the great weight of telescope, axis and circle.

By an agreement reached at the International Meridian Conference in Washington in 1884, the meridian defined by this instrument became accepted as the prime meridian of the world. The whole elaborate system of international longitude reference and international timekeeping, defined by the Greenwich Meridian and by Greenwich Mean Time, thus depended on this unique transit circle.

The equatorial

Pearson, as we have seen, was not enthusiastic about the equatorial as a positional instrument, and claimed to represent the view common among astronomers. Nonetheless, an equatorial telescope was responsible for one of the famous achievements of precision astronomy in the nineteenth century, the discovery of stellar parallax, which had defeated the efforts of astronomers and instrument-makers since the time of Tycho.

An equatorial after the design of Fraunhofer (fig. 200) consisted of a stand for an adjustable polar axis, with a right ascension circle (sometimes known as the hour-circle) at its lower end, where a clockwork drive was applied to its rotation. At the upper end of the polar axis was mounted the declination axis, with its circle at one end and the telescope at the other. Fitted with a divided object glass micrometer, this instrument became a heliometer, capable of accurately measuring small angular distances. The Königsberg Observatory in Prussia ordered such an instrument from Fraunhofer, and it was completed by Utzschneider in 1829. (They also, incidentally, purchased a transit circle from Reichenbach and Ertel in 1819, and another by Repsold in 1841.) It was with the heliometer that Friedrich Wilhelm Bessel (1784–1840) detected the parallax of the star 61 Cygni in 1838 – the first observation of stellar parallax and the first measurement of the distance of a star.

German heliometers were popular additions to observatory instruments during this period, and were installed in Bonn, Göttingen, Hamburg and Munich (Bogenhausen Observatory), and also in Budapest, Pulkowa, Warsaw, Breslau and Helsinki. Most commonly made by Utzschneider and Fraunhofer, they were also built by Reichenbach and by Merz. The single example in Britain (fig. 201), set up in the Radcliffe Observatory in Oxford in 1849, was by Repsold, with optics by Merz, one obvious difference from the Fraunhofer instrument being that the declination circle was placed at the upper end of the polar axis.

Fraunhofer's heliometer and the Greenwich transit circle in their different ways mark culminating achievements of the divided circle tradition in astronomy. The one, in 1838, met the great observational challenge of theoretical astronomy, set almost three hundred

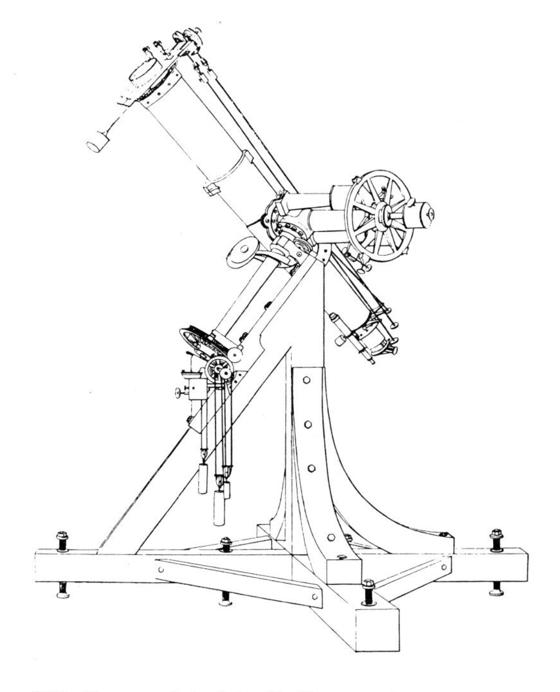

200 Equatorial of the 'heliometer' type, built by Fraunhofer in 1826 (Repsold, 1908)

201 Heliometer by Repsold, optics by Merz, 1849, focal length 3.2m, built for the Radcliffe Observatory, Oxford. The divided object-glass micrometer is clearly visible, as are the right ascension and declination circles. The weight hanging at the front is for driving the clock. London, Science Museum.

years before by Copernicus. On the other, in 1884, was based the ultimate achievement of a very practical technology – the world encased in a set of longitude meridians and time zones, accepted by international agreement and serviced by practical astronomy. Theoretical astronomy, however, was by this time moving rapidly in quite a different direction. Observatories in the second half of the century were commissioning large equatorial telescopes not intended for precision work. Even Airy had such an instrument, with an object glass by Merz, installed at Greenwich in 1859. Instruments of this type would in time be fitted with spectroscopes or built as giant astronomical cameras. The research edge of astronomy had moved to a different field, to create the domain of astrophysics, and to relegate the technology of the divided circle, for the most part at least, to the role of a practical necessity.

12

Reform of Navigational Practice

NINETEENTH-CENTURY NAVIGATIONAL techniques and instruments derived in large part from the achievements of the eighteenth century, now to be rationalized and given widespread application. There were two reliable methods for finding longitude, where before there had been none. Though often seen as rivals, the two methods were in some ways complementary. The chronometer method might be more accurate in the early stages of a voyage, but irregularities of rate would lead to greater inaccuracy as time passed. Lunars was generally less precise, but fairly constant, and it could therefore serve as a check on the chronometer. Both methods were practised in the nineteenth century, but in the long run the chronometer was preferred.

The instruments involved were also complementary. The navigator still needed to determine local time, by octant or sextant, for comparison with time by the chronometer. The chronometer also assisted the 'double altitude' method of checking a latitude estimate, where an observed altitude was taken and, given the ship's speed and an estimate of her latitude, a predicted altitude was calculated and checked after an interval timed by the chronometer. Again, by timing intervals between longitude determinations by lunar distances, the chronometer facilitated the averaging of a number of individual results.

With the development of 'position-line navigation' in the mid-nineteenth century, sextant and chronometer became even more closely allied. One method, published by the American Thomas H. Sumner in 1843, involved an observation of the Sun's altitude, and two calculations of the ship's longitude by the chronometer, using two assumed latitudes (assumed latitude and solar altitude yielding a value for local time). These two positions could be plotted in latitude and longitude on a Mercator chart and the ship had, at the time of observation, to be on the straight line joining the two, or its extension. A similar observation and calculation were made at a later time, and the intersection of the second line with the first, transferred to allow for the intervening run of the ship, gave her position at the second observation. One obvious advantage of this technique was that the navigator was no longer restricted to observations at or around noon. In time, the Sumner method was largely replaced by a

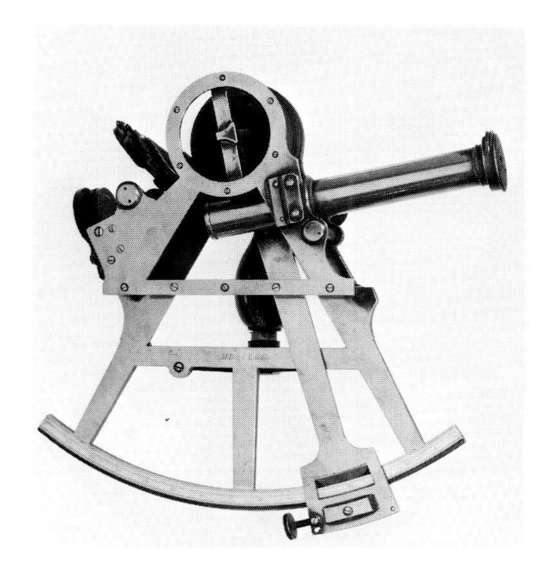

202 An early 19th-century example of Ramsden's bridge-frame sextant, by Berge, radius 184mm.

technique published in 1875 by the Frenchman Marcq de Saint Hilaire (1832–89).

With the longitude problem solved, instruments of astronomical navigation altered little in principle, in spite of a variety of technical experiments by the makers. Curiously, perhaps, more fundamental improvements were made to the kind of instruments used in the old bearing and distance technique. The magnetic compass at last received serious attention. Reliable mechanical logs were developed, since navigation involving solar or stellar sights depended on weather conditions and seamen still needed to measure speed or distance. Finally comprehensive changes were made to the earliest instrument of all, the lead and line.

Sextants and circles

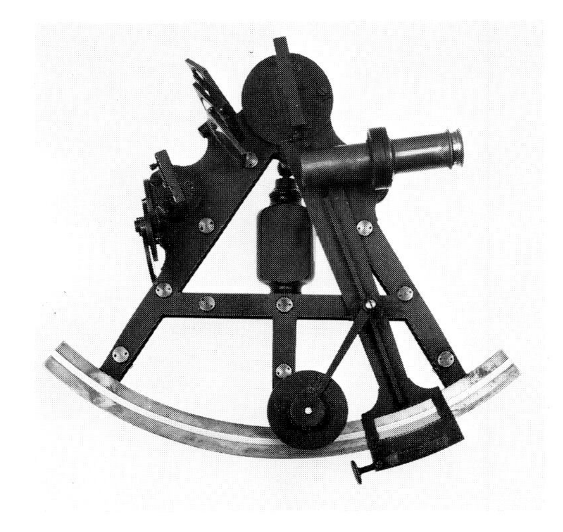

203 A 19th-century example of Troughton's double-frame sextant, by Troughton & Simms, radius 200mm, with an oxidized brass frame and platinum scale.

The nineteenth century saw no generally adopted changes in principle in the sextant. Smaller instruments became more common, settling down to a frame of radius around 20 cm, and there was much experimentation with alloys and designs for frames. Troughton's double-frame design (fig. 203) remained popular through much of the century, and there are examples by a good number of different makers. Indeed, Henry Hughes & Son Ltd advertised this model in their 1904 catalogue, with the note: 'We still make and supply a large number of this well-known pattern'. Ramsden's bridge frame (fig. 202) fared less well. It survived him mainly through the production of Berge – who styled himself 'late Ramsden' – but also of Thomas Jones, who had worked for Ramsden. Ramsden's lattice frame, on the other hand, did become popular and was reproduced, both close to its original form (fig. 206) and in modifications, throughout the century.

The most popular designs for sextant frames later in the nineteenth century were a modified lattice and the 'triple circle'. The former (fig. 207) had two principal radial members, in addition to the outer radii of the frame, with short struts linking them to the outer radii, and a concentric member linked by short struts to the

204 Three early 19th-century octants, from the left: signed by W. Weichert of Cardiff, radius 240mm; by Bleuler of London, radius 280mm; unsigned, radius 240mm. All have pin-hole sights, but only the instrument by Bleuler has the back horizon glass. The scale on this instrument is stamped with an anchor, indicating that it was engine-divided. Some such stamps have initials, such as 'IR' for Ramsden's engine. The most common divider's mark is 'SBR' for Spencer, Browning & Rust.

205 (*left*) Octant by Soulby of London, first half 19th century, radius 273mm, illustrating a frame with two vertical braces, which is much less common than the standard T frame.

206 (*right*) Sextant, signed 'Stancliffe' and 'Divided by J. Allan', *c.*1800, radius 287mm. James Allan had a particular interest in dividing and built a successful engine; Stancliffe was recommended by Maskelyne, as a maker of sextants 'thought fully equal to Ramsden's' (Bennett, 1983, A). The frame is close to the lattice form of Ramsden. Cambridge, Whipple Museum.

207 (*below left*) Sextant with a modified lattice frame by Henry Hughes, mid-19th century, radius 181mm.

208 (*below right*) Sextant with a 'bell' frame by Heath & Co, early 20th century, radius 178mm. The handle is pierced by a hole for mounting on a stand. A clip holds the index arm in place, with fine adjustment by an endless screw, but is readily released for larger motions.

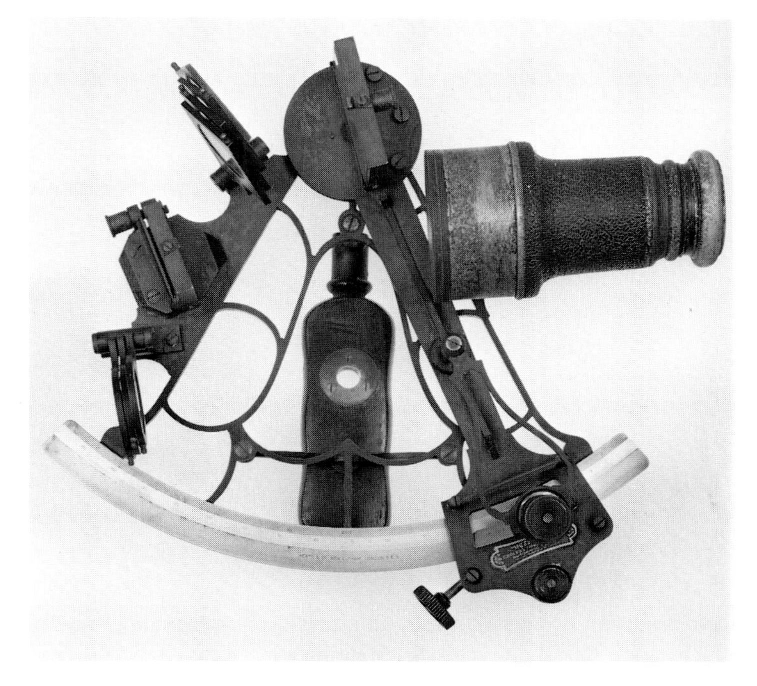

209 Sextant with a triple circle frame by Elliott Brothers, late 19th century, radius 165mm, with two alternative telescopes, one dark tube and two eyepiece filters. Cambridge, Whipple Museum.

210 (*below*) Double reflecting circle by Schmalcalder, *c.*1815, diameter 213mm. The two circles share the fixed optical components in the centre, but have separate index mirrors and quadruple index arms. The purpose of the instrument is not clear, since the double facility is usually associated with sounding surveying, but so sophisticated and accurate an instrument would be inappropriate. It is more likely that it allowed an observation to be made twice in quick succession, to combine some of the advantages of the Borda and Troughton circles. Whatever the answer, it did not become popular. Cambridge, Whipple Museum.

limb. This left a vacant central space, which facilitated handling. The triple circle (fig. 209) is more easily envisaged, with three circles enclosed by the outer frame. The T frame continued to be made, and a number of developments of the modified lattice frame were tried, often with quite elaborately curved members, and known as 'tracery' frames. One common design in this class was the 'Bell Pattern' sextant (fig. 208) made by Heath & Co. at the end of the century, so-called on account of the shape of the vacant central space.

If there were few changes in principle, there were innumerable variations in detail, as sextants came to be manufactured in great numbers and by many makers. An example of a reasonably typical sextant (fig. 209) from the late nineteenth century in the Whipple Museum, Cambridge, will have to stand for hundreds of other designs.

Signed on the limb 'ELLIOTT BROS. LONDON', it has a triple circle frame, a radius of 165 mm, and an index arm of 218 mm. The frame has been cast in brass and given an 'oxidized' finish. This is a dark grey-brown surface, induced chemically, which, by the second half of the century, was much more common than bright lacquered brass. The hardwood handle with finger indentations is set in an oxidized brass mount, attached by two pillars to the back of the frame, and is pierced by a bright brass fitting to allow the instrument to be mounted on a stand. Three brass feet at the corners of the frame project beyond the handle so that the instrument can rest on a flat surface.

The index arm, in oxidized brass, has been cast with a central reinforcement running along its length. The mount for the index mirror is adjustable, by three screws, and the mirror itself, held by three clamps, is readily replaced. Four pivoted shades, of different densities, for sighting the Sun, are mounted on the edge of the frame, just below the index mirror, and below them the replaceable half-silvered horizon glass. The horizon sight has three shades, pivoted

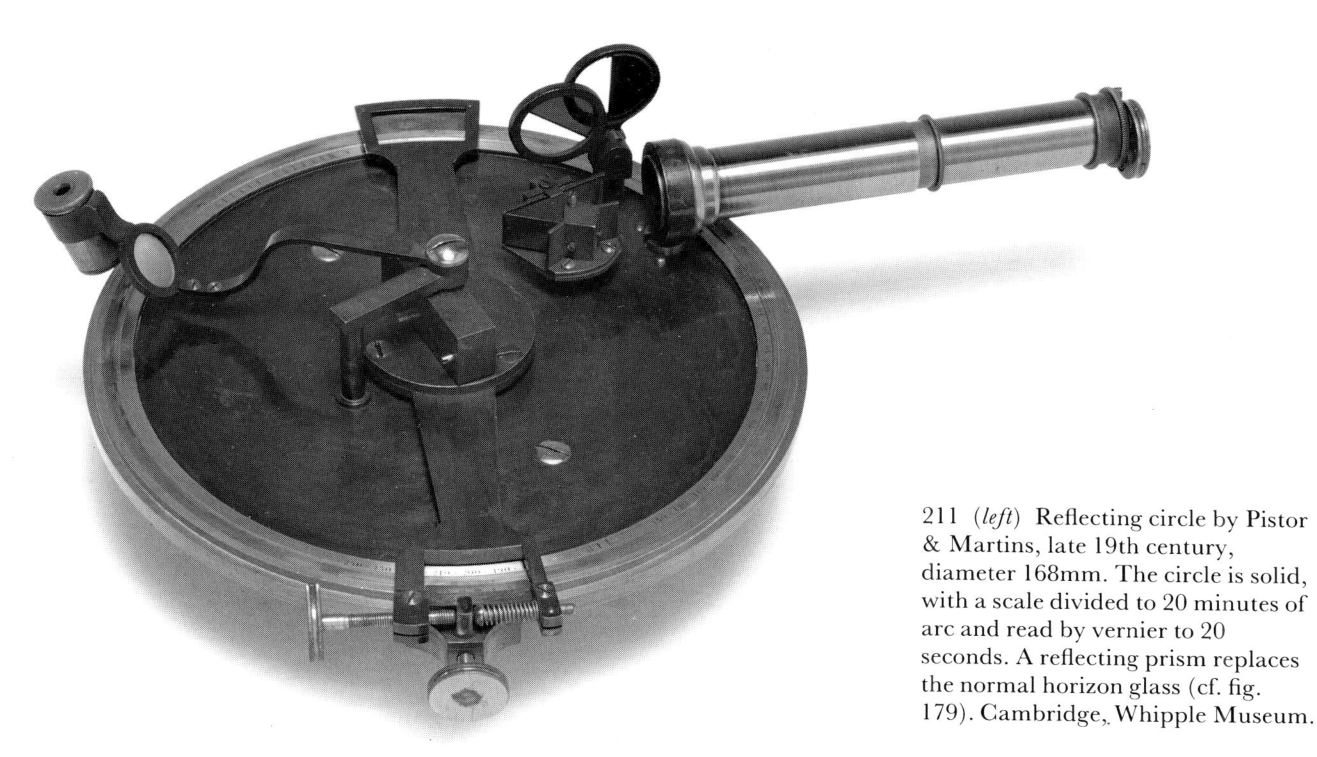

211 (*left*) Reflecting circle by Pistor & Martins, late 19th century, diameter 168mm. The circle is solid, with a scale divided to 20 minutes of arc and read by vernier to 20 seconds. A reflecting prism replaces the normal horizon glass (cf. fig. 179). Cambridge, Whipple Museum.

on a mount just below the glass. On the other radius of the frame is the mount for the telescope, which is adjusted laterally by a knurled screw projecting to the back of the frame. Two screw-fit telescopes are provided, of different focal lengths, and a third eyepiece, with an alternative arrangement of cross-hairs. There is also a 'dark tube', with only a pin-hole sight and no lenses. There are two screw-fit filters for the eye-ends of the telescopes or dark tube.

The brass limb, which is not oxidized, has an inset silver scale, divided from −5° to 160° and subdivided to ten minutes. It can be read by vernier to ten seconds. Fine adjustment of the index arm is by a clamp, applied by a knurled screw at the back of the limb, and a tangent screw moved by a knurled head placed on the far side from the observer. A push-focus reading microscope is attached to an arm which pivots in a mount on the index arm, and has a hinged, frosted glass reflector to illuminate the scale and vernier. The serial number '814' is engraved on the limb, and pasted in the box is a 'Class A' certificate from Kew Observatory, indicating the corrections to be applied to different portions of the scale, as detected during examination in June 1893.

212 Reflecting circle by Wanschaff, 1875, with a prismatic horizon 'glass' and a counterpoised mount on a stand. London, Science Museum.

Octants continued to be made throughout the nineteenth century with no important changes, except for the introduction of brass frames. Hughes & Son were still offering brass octants in 1904, but could only supply ebony ones second-hand. Circles remained more popular with continental than with British seamen. Troughton introduced a repeating form late in his career, but with little success. Secretan, on the other hand, was still manufacturing the Borda-type circle in the second half of the century. It was also included in the Hughes & Son 1904 catalogue, but it is difficult to imagine that there were many customers.

A relatively popular instrument (fig. 211) was introduced by Pistor & Martins, where a circle (solid in later examples) had a diametric index arm with verniers at both ends, and a right-angle reflecting prism replacing the horizon glass. The observer sighted over the prism for the direct ray and into it for the reflected ray. Pistor & Martins introduced reflecting prisms into other instruments, such as quintants, with arcs somewhat larger than the sextant.

213 Marine chronometer by Thomas Earnshaw, *c.*1803, diameter 60mm, gimbal-mounted in a mahogany box. The early history of this chronometer, serial number 610, is known, since it was in the possession of Sir Robert Stopford on board H.M.S. Spencer between 1803 and 1806, and Earnshaw used Stopford's certificate of its performance in his printed *Appeal to the public* of 1808.

214 (*below*) Movement of Earnshaw's chronometer no. 610 (fig. 213), showing the helical balance spring and bimetallic compensation balance with segmental weights.

The chronometer

John Arnold and Thomas Earnshaw were responsible for the earliest production of marine chronometers (figs 213, 214, 215) in large numbers and in a standard form. While Arnold turned out a greater number of instruments, it was Earnshaw who developed the customary features.

In the standard instrument of the nineteenth century, the 'movement', as the working mechanism is called, is housed in a brass bowl and slung in gimbals (after the manner of a marine compass), suspended in a mahogany box. Often bound with brass at the corners, the box usually has brass carrying handles, and often has a double lid: the top alone hinges back for observing the time through a glazed window, and the lid proper need only be opened for winding or adjusting.

Beneath a screw-fit glazed 'bezel', the dial is usually silvered, and commonly shows twelve hours. Some chronometers have 24–hour dials; usually designed for astronomical use, they were often rated to sidereal rather than mean solar time. The main dial shows hours and minutes, with seconds usually presented on a subsidiary dial. There is generally also an 'up and down' dial, showing how soon the chronometer will need rewinding. The dial will bear the maker's or the retailer's name and address, a serial number, and perhaps some description of the maker's standing, such as 'Maker to the Admiralty, the Indian and Italian Governments' and so on.

The movement itself is generally very fine and highly finished. Two circular brass plates are separated by brass pillars. The mainspring, fusee, train and escapement are contained between these plates, and above them (if the movement is examined with the dial down) is the balance and balance spring.

The mainspring is contained within a brass barrel, one end being attached to the barrel, the other to a steel rod known as the 'barrel arbor', which is equipped with a ratchet. Thus the barrel rotates as the force of the wound spring is released. With the chronometer unwound, the barrel has a steel chain – the 'fusee chain' – wound along its length and attached to it at one end. The force of the mainspring is transmitted through this chain and the nearby 'fusee', a brass cone cut with a spiral groove. The fusee arbor is square at its upper end, to take the winding key, and during winding the chain is transferred onto the fusee from the barrel, the barrel is turned and the mainspring coiled within.

The significance of the fusee is easily understood. The force of the mainspring is maximum when fully wound, and diminishes as it unwinds. For accurate timekeeping, however, a fairly constant force must be transmitted through the train to the balance, and the fusee has the effect of equalizing these variations. Thus the maximum force, when the spring is fully wound and the chain is at the apex of the fusee, is delivered closest to the fusee arbor, and the decreasing force is delivered further from the arbor as the spring unwinds. By the law of the lever, the force transmitted by the fusee arbor is thus fairly constant.

Inside the fusee is the 'maintaining power' – a mechanism whereby a subsidiary spring acts to maintain a force on the train, and keep the chronometer going, while the force of the mainspring is removed during winding. The design derives in principle from an invention of John Harrison. The 'train' – the gearing which transmits the force and operates the hands – has four brass cog wheels, as well as the 'escape wheel', and four steel pinions, which are small wheels with relatively few teeth or 'leaves'.

The final wheel of the train is the escape wheel, which forms part of the 'escapement'. The escapement and the balance have mutually related functions. The escape wheel acts on the balance to maintain its oscillation; the balance in turn, through the escapement, controls the release of the force at regular intervals and thus the timekeeping of the chronometer. The rest of the movement is carefully made and finished to high standards, but it is the escapement and balance that particularly characterize a chronometer. The term has been used more loosely of late, but traditionally a chronometer had a 'spring-detent' (or 'chronometer') escapement and a 'compensation' balance.

215 Marine chronometer by John Roger Arnold (son of John Arnold), c.1825, diameter 64mm.

Most chronometers after Earnshaw's were fitted with his form of escapement, the 'Earnshaw spring-detent' escapement. Essentially, a piece known as the detent (because it detains the escape wheel) is mounted on a spring, and for much of the oscillation of the balance locks the escape wheel. However, at a particular point, close to the centre of oscillation in one direction, a pallet attached to the balance staff moves the detent briefly to one side. This releases the escape wheel, one tooth of which acts on a second pallet (gives impulse to the balance, as horologists say), fixed to a 'roller', which is also mounted on the balance staff. By this time the detent has been returned by its spring and is ready to lock the escape wheel once again.

The advantage of this arrangement is that the escape wheel interferes as little as possible with the oscillation of the balance, and so with its isochronous character, impulse being given only for an instant, near the centre of oscillation, and always in the same direction. Other forms of escapement are known. Some French makers, most notably Jean-François-Henri Motel (1786–1859), used a detent which was pivoted rather than mounted on a spring. A number of more exotic escapements occur occasionally, but Earnshaw's is by far the most common.

The balance spring is generally a simple helix with specially formed 'terminal curves', which establish the isochronous character of the oscillation. The mathematical account of the theoretically correct forms for the terminal curves was laid down by the French mathematician Edouard Phillips (1821–89) in 1861. Blued steel is the most common material, though gold and palladium springs are also found. Edward John Dent (1790–1853) tried to use springs of glass. Various forms more complex than the simple helix occur very occasionally.

The balance is the component of the chronometer which has allowed most scope for experimentation. The basic compensation balance, in its common form, is also due to Earnshaw. The balance has a straight steel arm, fixed at its centre to the balance staff, and at either end are fixed curved bimetallic rims of brass and steel, each with a weight, positioned symmetrically. The idea is to compensate for the effects of changing temperature on the spring's elasticity. As the temperature rises, and the spring becomes less elastic, the period of oscillation would tend to lengthen and the chronometer run slow. The bimetallic rims of the balance, however, act to move the weights towards the centre, thus increasing the moment of inertia of the balance and effecting the compensation. As the temperature drops, the reverse applies. The basic balance has two adjustments. Screws at either end of the central arm may be adjusted to change the rate, and the positions of the weights may be changed, symmetrically, to alter the amount of compensation.

Although it is the most common form, the basic compensation balance does not achieve complete temperature compensation, and a great deal of inventive effort in the mid-nineteenth century was directed at devising improved balances. Thus many chronometers are described on their dials as having some 'auxiliary' compensation. This is usually some addition to the standard balance, and is generally 'discontinuous', acting only in certain extremes of heat or cold. Examples of discontinuous auxiliary compensations were those devised by the chronometer makers John Poole (1818–67), Thomas Mercer (1822–1900) and Victor Kullberg (1824–90). Other balances are quite different from the standard form, and provide continuous compensation. Designs of this sort were produced by the makers E. J. Dent and Edward Thomas Loseby (*fl.*1846–90), and by the astronomer John Hartnup. Loseby's is the most comprehensible, consisting as it does of mercurial thermometers, specially shaped and mounted on the ends of the balance rims, the motion of the mercury in the glass tubes providing the auxiliary compensation.

Another addition found in some balances was known as 'Airy's bar', after its inventor G. B. Airy. Here the ordinary balance has an additional bar, mounted friction-tight on the balance staff, and having short springs at both ends, with weights that press on the balance rims. As the position of the bar is changed, the weights press at different positions on the bimetallic rims, so that Airy's bar is merely a device for adjusting the amount of compensation without moving the main weights.

Two other forms of navigational timepiece were, firstly, the 'chronometer watch', a pocket-watch that conformed to the characteristics of a marine chronometer, having a chronometer escapement and a compensation balance. This was used for carrying time from the chronometer to wherever it was needed on the ship for navigation, since it was preferable not to move the chronometer itself. In port, the watch was used for checking the chronometer's time by a regulator clock. Secondly, a 'deck watch' was a good quality pocket-watch, with a similar function, but not meeting the requirements of a chronometer. It might, for example, have a lever escapement, the type of escapement used in most mechanical watches.

The compass

We have seen that a growing awareness of the inadequacies of the marine compass had suggested the adoption of a standard instrument that could be made in numbers and issued to ships. By the mid-nineteenth century, the British Admiralty's standard compass had a copper bowl and four straight steel needles, attached to a ring beneath a 7.5 in paper card. The idea had also emerged of a ship having a standard compass in a different sense, namely one master compass of a superior design and in the best possible position. This was used to regulate the others, such as the steering compass at the ship's wheel. The azimuth compass had previously been used at any convenient position on the ship, but the stationary standard compass was now equipped with sights and used for taking bearings. The Admiralty standard compass of the mid-nineteenth century was used in this way, and long after it had been superseded as the ship's standard compass, it was carried as an auxiliary for taking magnetic bearings on land.

The 'prismatic sight' was invented by the London instrument-maker Charles Schmalcalder in 1812. The near-sight incorporated a right-angle prism, which reflected an image of the stationary index and part of the scale into the observer's eye. By modifying the shape of the prism, the image could be magnified. It was later found that the reflecting prism could alternatively be incorporated into the far-sight.

By the end of the nineteenth century, there were two distinct types of compass, the liquid compass and the dry-card compass. Both became popular, though among different clientèle, and in time the modern magnetic compass evolved through adopting the best features of both instruments.

The standard type of dry-card compass arose from a study of the problems caused by magnetic deviation – the disturbing influence of ferrous metal in the neighbourhood of the instrument. These difficulties became more acute as metal was increasingly used in shipbuilding and fitting. Investigations were carried out, most notably by the French mathematical physicist Simeon-Denis Poisson (1781–1840), and by G. B. Airy. Airy, who began his work in 1839, identified the different effects of variable magnetism induced in the soft iron parts of a ship by the Earth's magnetic field, and the permanent magnetism hammered into the ship during construction. He suggested correcting the permanent magnetism by introducing counteracting permanent magnets, and controlling the induced magnetism by soft iron masses placed close to the compass.

These ideas were developed by Sir William Thomson (later Lord Kelvin) (1824–1907), and two forms of his dry-card compass and binnacle (fig. 218) were patented in 1876 and 1879. The binnacle

216 Marine compass by Dent, *c.*1850, diameter 214mm. Dent was better known as a clock and chronometer maker, but he also designed and made compasses. This one has a vertical pivot resting in a bearing at the bottom of the bowl. The pivot is fixed to the card, beneath which are suspended four magnets and above four weights for adjusting it level. The degree scale is printed in reverse, indicating the use of a prismatic or other reflecting sight. Cambridge, Whipple Museum.

was designed so that all the correcting magnets could be attached to it – adjustable permanent magnets below to neutralize the ship's permanent magnetism, and movable soft iron spheres to either side to correct the induced magnetism. A third mass of correcting soft iron was a vertical bar housed in a cylindrical brass case screwed to the outside of the binnacle; this was known as the 'Flinders bar', and was due to the earlier work of the hydrographer Matthew Flinders (1774–1814). Its purpose was to correct for the magnetism induced in the ship's ferrous metal by the vertical component of the Earth's magnetic field, which varies with latitude. In the Thomson binnacle, the Flinder's bar was sectional, and could therefore be adjusted to the appropriate length.

In addition to the design of binnacle, other revolutionary features of the Thomson compass were the card and needles (fig. 217). A 10 in. diameter aluminium ring, with a paper ring printed with points and degrees cemented to it, was joined by thirty-two radial silk threads to a hub consisting of a sapphire cup with a metal cap. The threads also supported eight short steel needles. The idea of using short needles was to lessen the disturbing influence of the magnetism they might induce in the correcting soft iron spheres. The pivot was a brass rod with an iridium point, and the copper bowl had a chamber for weighting the compass with oil.

217 Compass card after Thomson's design by Kelvin & James White Ltd of Glasgow, *c.*1900, diameter 254mm. Four needles on each side of the pivot are held to the rim by a network of silk threads. Greenwich, National Maritime Museum.

Thomson's instrument was intended to be a standard compass and was naturally accompanied by a sight for taking azimuths. The device was called an 'azimuth mirror', because a mirror featured in the original form, but this was soon replaced by a 60–degree prism. The prism was carried above the compass card by an inclined tube (the range was thus not restricted by the correcting spheres), and sights could be taken in one of two alternative configurations of the prism. In one, an image of the card was seen in conjunction with an object sighted directly; in the other, the card was viewed directly, along with an image of the Sun or a star.

This compass was adopted by the British Navy, and was also widely used on British merchant ships. A great many later binnacles were also derived from Thomson's ideas. Another important development, however, had been proceeding at the same time, namely the practical improvement of the liquid compass. The liquid compass is so called because the bowl is largely filled with a liquid, generally a mixture of alcohol and water, the main advantage being a more steady card. The buoyancy also reduces friction and consequent wear on the pivot and bearing.

After a prehistory of suggestions and experiments, liquid compasses appeared in the first half of the nineteenth century, when some early makers were Francis Crow (*fl.*1813–32) and Grant Preston (*fl.*1832) in England, and Johan P. Weilback in Copenhagen. The instrument became a commercial reality in the mid-nineteenth century, around which time the Royal Navy introduced a liquid compass, of a type designed by the chronometer-maker Dent, for use in rough weather. But the most important commercial success was in America, where the instruments of Edward Samuel Richie (1814–95) of Boston were adopted by the United States

218 Compass and binnacle to the design of Sir William Thomson (later Lord Kelvin), by James White of Glasgow, *c.*1879, diameter 10in (254mm). The soft iron spheres to either side correct for magnetism induced in the ship by the Earth's magnetic field; the open door reveals permanent magnets which balance the permanent magnetism of the ship. London, Science Museum.

Navy. The needles of Richie's early compasses were contained inside a float, which carried the card. The principal feature of the float he introduced in the 1860s was a tubular ring, while in 1882 he devised an H-shaped float, with a needle in each of the parallel strokes of the H. In the early years of the twentieth century, he introduced a hemispherical float, with the domed side facing upwards, and this became widely adopted. Liquid compasses became common also in the French and Italian navies.

The pelorus, the azimuth circle and the deflector are three important accessories to the magnetic compass. Also known as the 'dumb card' or 'bearing plate', the pelorus was used for taking magnetic bearings of objects obscured from the standard compass. Although engraved like a compass card, it had no magnets, and its measurements were derived from those of the standard compass. It consisted of a brass plate, engraved with points and degrees, with sights that could rotate about the centre. The plate rotated within a ring

221 (*opposite, top*) Mechanical sounder by Edward Massey, early 19th century, 125 by 212mm, detatched from the lead, with a box and instruction label. A worm screw moves two dials, 0–10 fathoms, divided to 0.5 fathoms, and 0–150 divided to 5. The plate above the rotator locks the fins while the instrument is hauled in. Cambridge, Whipple Museum.

222 (*opposite, below*) A Massey sounder of the type illustrated in fig. 221 attached to a lead. London, Science Museum.

marked with a lubber's point, and the whole was slung in gimbals in a wooden box. The lubber's point had to be aligned with the fore-and-aft line of the ship, and the plate set so that the point coincided with the course being steered.

An azimuth circle was used for taking bearings with the magnetic compass. The term was used in a general sense, and a number of different optical arrangements were tried. The components were sometimes those of the 'azimuth mirror' used with the Thomson compass, but made up as a separate instrument, with the parts attached to a non-magnetic ring that fitted over the compass bowl. Other combinations of sights, mirrors and prisms were used; in some arrangements, for example, the Sun's bearing is indicated by a spot of light reflected onto the card.

The deflector was yet another instrument devised by Thomson, and was used to check for magnetic deviation at times when weather conditions precluded astronomical sights. The method depended on the introduction of an adjustable magnetic moment, whose variations could be measured, and the instrument consisted of a hinged pair of magnets, the distance between a pair of opposite poles being adjustable by a screw with two contrary threads and measurable by a micrometer.

Mechanical logs

We have seen that the history of the hand log is fairly uneventful, but the mechanical log – often known as a 'patent' log – has both a complex history and a long prehistory, with suggestions going back to Vitruvius. One was devised by Humfrey Cole in the sixteenth century, and seventeenth-century inventors included Hooke and Wren. Designs abound in the eighteenth century; some were patented, but none was commercially successful. In 1772 William Foxon of London patented a mechanical log (fig. 140), which consisted of a helical rotator, towed behind the ship, and linked by the log-line to an inboard movement. One dial recorded the distance run, and another was used in conjunction with a half-minute glass to measure the ship's speed.

The first commercially successful mechanical log was patented in 1802 by Edward Massey (*fl.*1802–36). His earliest model had two

219 (*left*) 'Harpoon' log by Thomas Walker, *c.*1865, length 460mm. Three dials record miles, tens and hundreds of miles, and a cover is rotated to enclose the dials before the log is cast astern. Cambridge, Whipple Museum.

220 (*right*) Taffrail log by John Bliss & Co, *c.*1900, length (register) 190mm. The register, pivoting in a fork mount, has dials for miles, tens and hundreds of miles. Cambridge, Whipple Museum.

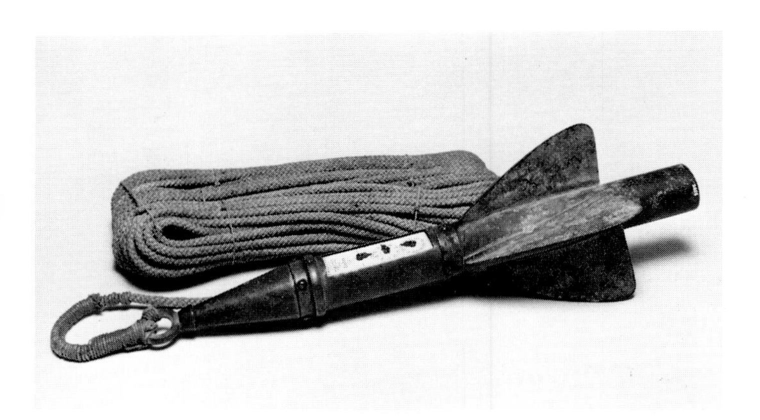

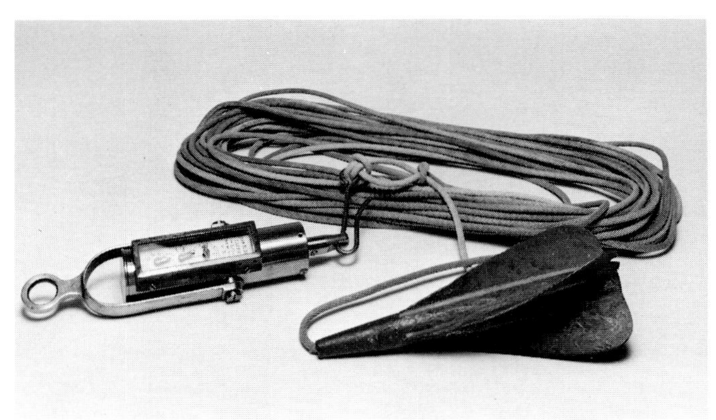

units: the 'rotator' or 'fly' – a pointed metal tube with vanes – was linked by a jointed cane to a geared counting mechanism with three dials, showing tenths of miles, miles and hundreds of miles, and the whole was towed by the log-line. The cane was later replaced by a rope.

During the nineteenth century, Massey and his successors took out a number of patents for improved logs. Massey's 'frictionless log', for example, was patented in 1865, and here the rotator and register were together in a single unit, with the outside rotating and the central section attached to the log-line. A later Massey log, by John Edward Massey, had the two functions again separate, but now with the register fixed at the stern of the ship.

Edward Massey's nephew Thomas Walker (1805–73) had a successful business in mechanical logs. Walker's 'harpoon log' of 1861 was a similar, but earlier, combination of rotating and recording functions, along the lines of Massey's 'frictionless'. In 1878 his son Thomas Ferdinand Walker (d.1921) also moved on to a 'taffrail' log, with a towed rotator and a register mounted at the stern. The advantage of the taffrail log was that it did not need to be hauled in for every reading. Walker's logs became very popular; they were especially widely used in the British Navy, and successors to the 'harpoon' were named 'Neptune', 'Cherubal', 'Cherub' and 'Excelsior'. Chester Gould (*fl.*1794–1803) of Boston was an early maker in America, and John Bliss (*fl.*1840–78) of New York produced a popular taffrail log (fig. 220), as well as a log which doubled as a mechanical sounder.

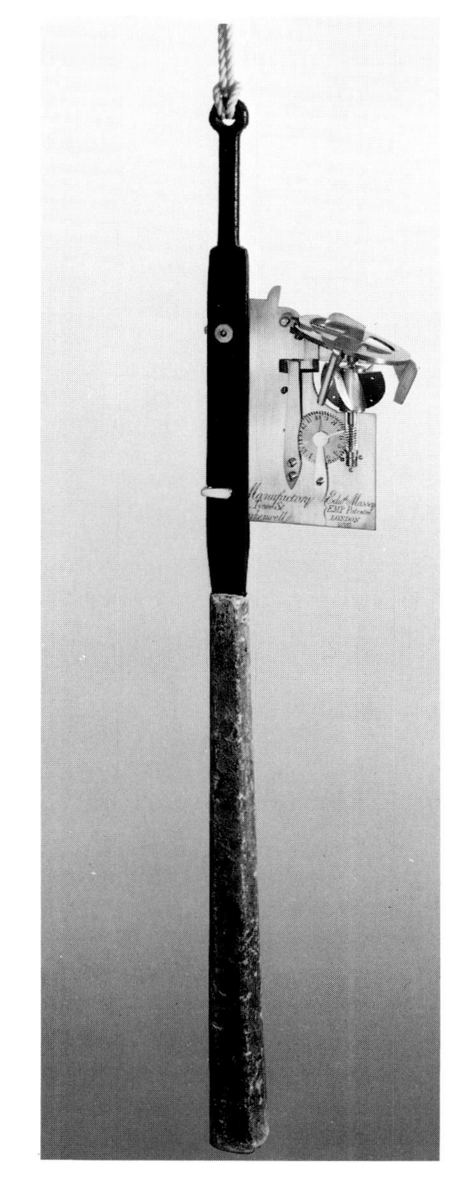

Mechanical sounders

Edward Massey was first again to manufacture a mechanical sounder, patented in 1802. It consisted of a rotator – a cylinder with shaped fins – and a counting mechanism, all attached directly to a long lead. After modifications and improvements Massey's sounder (figs 221, 222) was a qualified success. A device was introduced to lock the rotator when hauling in, as was a guard to protect the fins, the most vulnerable part of the instrument. A 'frictionless' sounder fig. 223) was later introduced and, unlike the early model, it was quite separate from the lead, and was towed behind it by a rope.

Other makers, such as Walker and E.& G. W. Blunt, followed Massey with sounders similar to the 'frictionless'. Le Coentre in France, on the other hand, developed a design where the register was again fixed to the lead, but operated when the machine was being hauled in. The main advantage of the Massey type of sounder was that the ship did not need to stop for sounding, provided the line was played out so as not to check the machine's vertical descent.

The same advantage applied to a second type of sounder, called a 'buoy and nipper', where the line ran through a spring catch on a buoy cast with the lead. The catch locked the line, thus marking the depth, when the lead hit the bottom. This device was sometimes called a 'bag and nipper' because of the canvas buoy or bag cast with the lead.

A more straightforward device, used during the nineteenth and early twentieth centuries, was the 'sentinel' or 'submarine sentry'. A kite-shaped board was towed astern by a wire and, rather as a kite is elevated by the wind, the oblique position of the sentinel caused it to submerge to a depth indicated by a dial plate. If it touched bottom, it was released by a trigger, rose to the surface and could be seen in the ship's wake. Simultaneously, it sounded a gong on board. In later models, an electric circuit was completed when the sentinel touched bottom, and a bell would sound on the ship.

An instrument patented in 1835 by John Ericsson of America operated on quite a different principle from the mechanical sounders, by relating depth to the maximum pressure encountered in a descent to the bottom. The gauge consisted of a glass tube with a restriction at its open end (early models had a valve), and the length of water, forced in against the pressure of the trapped air, was a measure of depth. With such an instrument, the path of descent – critical to the mechanical sounder – was immaterial.

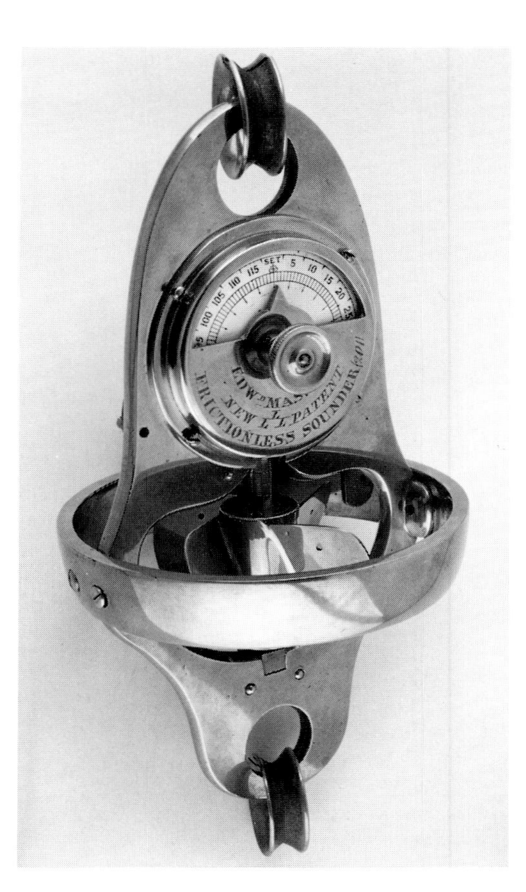

223 'Frictionless' sounder by Massey, *c.*1830, with the fly visible just beneath the dial. London, Science Museum.

The pressure principle was employed in the sounder invented by Sir William Thomson (fig. 225) and patented in 1876. It could take accurate soundings of up to 150 fathoms with the ship travelling at full speed. A line of pianoforte wire was wound on a drum, secured to the deck at the stern of the ship. A brake locked the drum when the sounder reached the bottom and the wire slackened. At the end of the wire was a cord, attached to which was a galvanized iron weight of 22lbs (10kg) and a brass cylinder containing the glass sounding tube. The inner surface of the glass tube was coated with orange silver chromate, and as sea water was forced into the tube, it reacted with the silver chromate to produce white, insoluble silver chloride. After retrieval, it remained only to place the tube against a specially calibrated boxwood scale and read the length of discoloration.

In early models, a proper measure of depth came only from the tube, but since a dial showed the length of wire let out, a rough estimate – or at least an early warning of danger – was available to an experienced operator. The skill here involved making allowance for the ship's speed, and carefully applying the brake used to control the release of the wire. Some later machines had an automatic brake, and came with a table relating length of wire to depth for different speeds.

Thomson's sounder was popular, becoming standard in the British Navy, and it was imitated. The Tanner-Bilsch machine, for example, was more often found in American ships, especially those of the United States Navy. There was no chemical action in this design, but the inside of the glass tube was ground and became clear when wet. In Basnett's sounding machine, the water forced in was retained and its length measured. In Cooper and Wigzell's 'sea-sounder', patented in 1890, the water moved a piston against a spring and a pointer indicated the reading. These are some of the variants that appeared as makers sought to make use of the ideas incorporated in Thomson's sounder, as well as in his compass.

Squire Thornton Stratford Lecky (1838–1902), whose famous textbook *'Wrinkles' in practical navigation* ran through many editions in

224 Thomson with his binnacle and azimuth mirror. (Thompson, 1910)

225 Thomson's sounding machine, as illustrated in the 1904 catalogue of Henry Hughes & Son Ltd.

the late nineteenth and early twentieth centuries, wrote of the then Lord Kelvin:

> By turning his brilliant scientific abilities into practical channels – a rare thing with the higher order of mathematical minds – he has, beyond possibility of question, done more than any other living man to advance the science of Practical Navigation. He may justly be regarded with veneration as the great 'Sea-Father' of mariners of the present age.
> (Lecky, 1925)

Scientists had always been more interested in astronomical navigation, in the application of the divided circle to the needs of mariners. It was only in the final decades of pre-electronic navigation that a scientist had successfully made it his business to improve comprehensively the most ancient tools of sounder and compass.

13

New Standards in Surveying Practice

IN TERMS OF BOTH DEMAND AND PRODUCtion, surveying instruments probably benefited more directly than those of astronomy or navigation, from the development of industrialized societies. While national surveys established new standards of practice and instrumentation, social changes created unprecedented demand for practical surveying, through urban expansion and the development of roads, canals and railways. At the same time, mine and colonial surveying, with their special needs, created particular classes of instruments. On the production side, industrial methods allowed makers to respond effectively to this increased demand, and resulted in a considerable degree of standardization.

The altazimuth theodolite was finally to become generally accepted as the surveyor's principal badge of office, and by the end of the century complex adaptations of the theodolite were being applied to specialized work in civil engineering. By the mid-century, the theodolite's more humble cousin, the level, was evolving a standard form. The circumferentor was shunted more firmly into tasks for which it was particularly suited, but there it blossomed into more ambitious designs.

226 (*left*) Simple theodolite by Bradford of London, *c.*1820, diameter 203mm. An example of the design described by Adams in 1791 as the 'common theodolite', with a pair of fixed sights and a pair on the alidade. The protractor was an original accessory. Cambridge, Whipple Museum.

227 (*right*) Gunter surveyor's chain by Chesterman of Sheffield, 19th century, length 66 ft (20m), comprising 100 8in links, with brass tags or 'tellers' every tenth link. Cambridge, Whipple Museum.

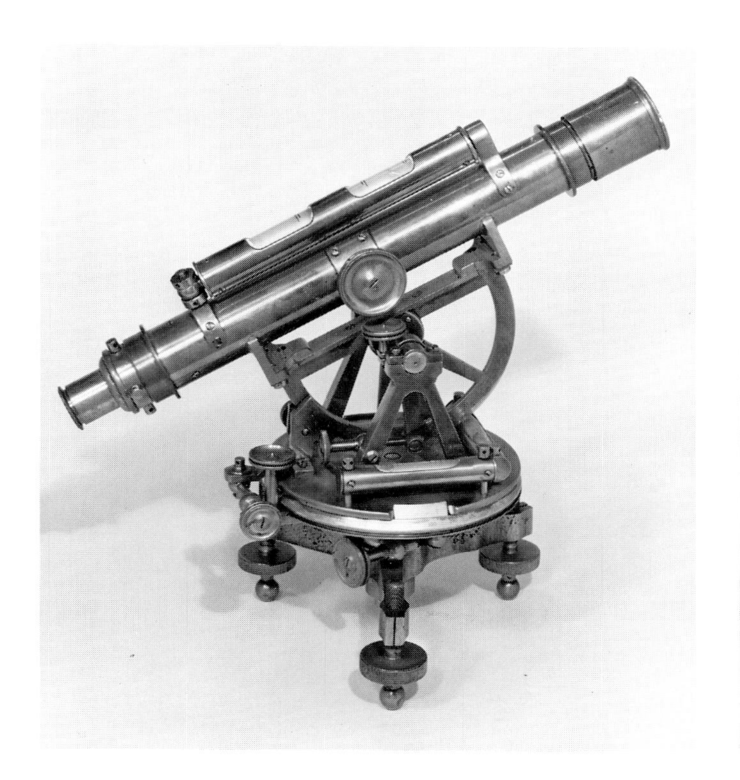

228 (*left*) Plain theodolite by Troughton & Simms, 19th century. This instrument has a flat tripod or 'tribrach' base, instead of the alternative parallel plates sandwiching four levelling screws.

229 (*right*) Everest theodolite by Troughton & Simms, *c.*1880, diameter 215mm. The horizontal circle is divided on silver to 10 minutes of arc and read by vernier to 10 seconds, and the similar divisions of the vertical arcs extend to 40° above and below the horizontal. The latter also has a 'Diff of Hypo & Base' scale (see caption to fig. 79). Cambridge, Whipple Museum.

Theodolites

By the time F. W. Simms wrote his *Treatise on the principal mathematical instruments employed in surveying, levelling and astronomy*, published in 1834, his representative theodolite was one very similar to the best instrument Adams had described in detail in 1791 and ascribed to Ramsden (see Chapter 9). This form (cf. figs 152, 228) was to become known as the 'plain' (sometimes 'cradle'), to distinguish it from the later 'transit', theodolite.

Two parallel plates with four levelling screws were used to set the vertical axis, which in turn supported two close-fitting horizontal plates, whose rims formed a continuous bevelled edge. The rim of the lower plate was divided to thirty minutes; that of the upper carried two verniers for reading to one minute. The upper plate, as well as a compass and two bubble levels, carried the two A supports of the horizontal axis. The two halves of the axis met at the centre of the vertical semicircle, which extended to a clamp and tangent screw fitting on the upper horizontal plate, with a vernier which again permitted readings to one minute. On top of the vertical semicircle were mounted the Y supports for the telescope, which had a hanging level and rack and pinion focusing.

Simms referred to the two most apparent omissions from Ramsden's design. The vertical arc was, he said, sometimes racked and moved by a pinion, 'but this is generally inferior, where delicacy is required' (Simms, 1834). (The horizontal motion, however, was usually by rack and pinion.) A second telescope was sometimes introduced beneath the azimuth circle, to check that the theodolite has not moved during the observation.

Simms also described the 'Everest' theodolite (fig. 229), designed by Sir George Everest (1790–1866) for surveying in India. This form

230 (*right*) Geodetic theodolite or portable altazimuth instrument by Troughton, 1806, diameter (horizontal circle) 370mm. Note the upward pointing conical bearing for the horizontal circle, and the familiar double-cone horizontal axis and double vertical circle with separating pillars. Each circle is divided to 5 minutes of arc and read by two micrometer microscopes to single seconds. This instrument was presented to John Playfair by his students at Edinburgh University, and was used by Charles Piazzi Smyth to survey the Great Pyramid in 1865 Edinburgh, Royal Museum of Scotland.

231 (*below*) Theodolite by J. G. Studer of Freiberg, early 19th century, but similar to Sisson's much earlier designs. Munich, Deutsches Museum.

generally had a 'tribrach' base, comprising three flat arms extending to levelling screws, an arrangement closer to contemporary astronomical instruments. Located in this base was the bearing for the vertical axis, which rose to the Y supports for the horizontal axis, in the manner of a transit instrument. The azimuth circle was fixed to the base, and four arms extended to it from the vertical axis – three indexes with vernier scales, and one arm with a clamp and tangent screw.

The telescope was also modelled on the transit instrument and was therefore fixed centrally between the double cones of the horizontal axis. Two short vertical arcs were fixed to the horizontal axis and further secured towards either end of the telescope. As they moved with the telescope, each scale could be read by a vernier attached to either end of an index arm that pivoted on the horizontal

232 Geodetic theodolite by Troughton & Simms, 1828, diameter 2 ft (610 mm), with a series of micrometer microscopes for reading the horizontal circle, emphasizing the importance of azimuth measurement. London, Science Museum.

233 Transit theodolite by Berge and Jones, early 19th century, diameter (horizontal circle) 176mm. A very early example of a transit theodolite, signed by two pupils of Ramsden: by Jones on the vertical circle and by Berge on the telescope. It may have been a collaboration, or a modification by one maker of the other's work. Cambridge, Whipple Museum.

axis and was fitted with a bubble level. This index arm could be clamped and levelled by two opposing screws that located onto one of two projections on the mount for the horizontal axis. When clamped and levelled, the index arm established the zero or horizontal line, and remained fixed while the telescope and scales were adjusted to the target, the final adjustment being by clamp and tangent screw. The whole assembly of horizontal axis, vertical scales and index arm could be reversed in the Y bearings, which had a fixed hanging level.

Everest's design was a bold attempt to model a radically new theodolite on the best principles of contemporary astronomical instruments. Some features became part of the dominant design of the second half of the century in an instrument somewhat confusingly called a 'transit' theodolite (figs 233, 234, 235). Simms, for example,

234 Two transit theodolites by Troughton & Simms: left, *c.*1905, diameter (horizontal circle) 90mm, '3 in'; right, *c.*1905, 119mm, '4 in'. Two typical examples of a very popular instrument. The detachable trough compass is conspicuous beneath the horizontal circle of the 4–in instrument. Cambridge, Whipple Museum.

pointed to the advantage of being able to determine altitude as a mean of two readings. The Everest theodolite, however, was modelled as far as possible on the transit instrument in astronomy, while the transit theodolite was more closely related to the altazimuth instrument, combined with features of the plain theodolite.

The term 'transit', applied to a theodolite, is in no way connected with a star's transit, referred to in the case of an astronomical transit instrument, but indicates rather that the telescope can be rotated about the horizontal axis so as to face in either direction – it can 'transit' across the A frame supports. The coincidence of terminology is unfortunate, but completely entrenched in the literature.

The transit theodolite is similar to the plain up to the horizontal axis, except that the A frame needs to be taller to allow the telescope to clear the base plate. The remainder of the instrument is more closely modelled on the Everest theodolite, except that the short vertical arcs have been replaced by a complete circle. This circle is mounted centrally with the horizontal axis and moves with the telescope as before, an index arm with two verniers being levelled by a striding level. The index arm is attached to a vertical 'clipping arm', at the end of which are two opposed screws (or one screw working against a spring); these act on a projection on the inside of each A standard and are used to set the index level. The clipping arm also carries the clamp and tangent screws for the fine adjustment of the telescope and circle. As with the Everest theodolite, the horizontal axis, circle and index can be removed as a whole and reversed in the Y bearings.

The transit theodolite first appeared in the 1840s, and in 1868 was 'now the favourite instrument' according to W. D. Haskoll (Haskoll, 1868, p.81). By 1890, W. F. Stanley could claim that 'The plain theodolite appears to be going gradually out of use, being superseded by the transit' (Stanley, 1890, p.236). While the numbers of surviving instruments confirm this trend, plain theodolites continued to be made well into the twentieth century. The transit had meanwhile expanded in range and was made in sizes from 3 to 12 in (the size is usually given as the diameter of the horizontal circle).

Towards the end of the century, circular compasses mounted on the horizontal plate were giving way to 'trough' compasses. In these, the needle had a range of only 10° or so on either side of magnetic north, and they came as accessories in the form of glazed rectangular boxes which could be mounted beneath the horizontal plate. Other standard accessories were plumb-lines and striding levels for the horizontal axis. The most radical change late in the century was Stanley's introduction of the U frame (fig. 235), a solid, single casting which replaced the two A frame standards.

In France, more theodolite designs were related in principle to the Everest pattern, in that the vertical axis was continued right up to the horizontal axis, and not simply to the horizontal plate as in the plain theodolite. The horizontal axis might then be carried in Y bearings, like a transit instrument or Everest theodolite, although other arrangements had the telescope and vertical circle mounted to one

235 (*above*) Transit theodolite by Stanley, *c.*1926, diameter (horizontal circle) 120mm, '4 in'. An example of Stanley's U frame where the support for the horizontal axis was cast in one piece, and the central compass dispensed with. Cambridge, Whipple Museum.

236 (*left*) Transit theodolite by Cary, 1915, diameter (horizontal circle) 145 mm, '5 in'. Fitted to the eye end is a 'prismatic astrolabe', a device invented at the beginning of the century for finding the latitude and longitude of an observing station. These are computed from timed observations of stars, noted when two elements of a double image – one after a single reflection in the prism, the other after reflections in the mercury tray and prism – coincide, when the star is at an altitude of precisely 60°. Note also the lamp and two micrometer microscopes for each circle. Cambridge, Whipple Museum.

side of the vertical axis. Later in the century, however, the transit theodolite also became familiar among instruments manufactured in France. The second telescope beneath the horizontal plate was more common in French than in English instruments.

Levels

The first edition of Simms's *Treatise on the principal mathematical instruments* of 1834 appeared shortly before the development of the standard model of theodolite. The situation is closely paralleled in the case of the level. Thus Simms described the plain theodolite, which owed much to Sisson, and the newly designed instrument of Everest. In treating levels, he dealt with the Y, due to Sisson, and the 'improved' model of Edward Troughton (fig. 238). The Everest theodolite and Troughton's improved level are both somewhat exotic instruments, the one having been overtaken by the transit theodolite and the other by the 'dumpy' level.

Troughton's improved level was an attempt to design an instrument whose basic adjustments could be fixed by the maker, in order to lessen the duties and responsibilities of the practitioner. This was an indication of the increasing demand for surveyors and the need for a less demanding instrument than the Y level. The familiar arrangement of four levelling screws between parallel plates carried a flat platform, on which the telescope was mounted with no provision for further adjustment. Above the telescope was the bubble level, again with no adjustments, since it was partially embedded in the telescope tube. Stanley pointed out that this was going too far, as bubble tubes were liable to be broken and there was no way of adjusting a replacement tube to the axis of the telescope. Above the level, carried by four pillars rising from the platform, was a circular compass, engraved with a compass rose and a degree scale; the level could thus be used as a circumferentor.

This level was not widely adopted, and examples are found only signed by either Troughton or Troughton and Simms. The dangers created for the historian by the frequent reprinting of illustrations in successive textbooks, are indicated by Stanley, writing in 1890: 'This

237 (*left*) Dumpy level by Aidie of London, *c.*1845, length 310mm, inscribed on the bubble level, 'Manchester and Leeds Railway Company'. The second bubble level is circular, instead of the usual short transverse form. Cambridge, Whipple Museum.

238 (*right*) An example of Troughton's 'improved' level, by Troughton & Simms, *c.*1845, length 357mm. Cambridge, Whipple Museum.

239 Three 19th-century 'primitive' levels: left, by Watkins & Hill of London, early 19th century, length 267mm, with plain sights; centre, by Huntley of London, *c.*1825, with a compass; right, by Gardner & Co. of Glasgow, c.1850, length 328mm, a drainage level. Cambridge, Whipple Museum.

level has gone out of use, but is mentioned here as the engraving of it remains in our *modern* text-books' (Stanley, 1890, p.110, emphasis as in the original).

The advantages of Troughton's level were shortly to be supplied by an instrument superior on several other counts and designed by the civil engineer William Gravatt (1806–66). Sometimes known as 'Gravatt's level', but more commonly as the 'dumpy level' (fig. 237), the familiar name indicates the most obvious change. Gravatt took advantage of the increased reliability of manufactured object lenses to allow a larger aperture and a shorter focal length. The result was a more compact and lighter instrument, with a wider field of view. The ability of the opticians to centre the optical axis of the telescope accurately was another pre-requisite for minimizing adjustments by dispensing with the telescope levelling screw of the Y level, and the dumpy was similar to Troughton's level up to the telescope tube.

The bubble tube was carried above the telescope, in a mount that was not adjusted in ordinary use, but which allowed replacement and resetting as required. An additional, short, transverse bubble, was regarded by Stanley as 'one of the most important structural improvements' (Stanley, 1890, pp. 110–11), since with it most of the setting-up procedure could be completed without rotating the telescope. If a compass was included in the dumpy level, it was incorporated into the base plate, sometimes with a reflecting prism for reading a pivoted card, like the prismatic compass (see below).

Stanley surmised that Gravatt '. . . found the less open adjustments the better in the hands of the imperfectly trained assistants who were pressed into service during the railway mania' (Stanley, 1890, p.110), but in fact his design gained widespread popularity, especially in England. Continental levels tended to have more adjustments to the telescope mount, though the telescope itself was not always carried in Ys. Meanwhile, the Y level survived, but its telescope took on the form of the dumpy telescope. As with the theodolite, there was a gradual tendency later in the century to replace the arrangement of parallel plates and four levelling screws, with three screws and a 'tribrach' (flat tripod) base.

240 On the left is a simple reflecting level, by Stanley, *c.*1900, length 122mm. The instrument is suspended vertically in front of the eye, when the distant object that coincides with the reflection of the eye's pupil in the half mirror will be at eye level. On the right is 'De Lisle's reflecting clinometer', unsigned, *c.*1920, where the same principle is extended to measuring inclines, by adding an adjustable weight which swings the suspended mirror out of the vertical. Cambridge, Whipple Museum.

Mine surveying instruments

A standard circumferentor is sometimes referred to as a 'miner's dial', and though the name does not do justice to the instrument's range of application, it is not inappropriate. During the nineteenth century, the circumferentor in its simple form gradually ceased to be used for general survey work, but was adapted rather to specialist applications in mine and colonial survey. From the surveying point of view, there were important similarities between the two sets of conditions. In a virgin territory, as well as down a mine, the surveyor was liable to find himself fixing directions without clearly defined landmarks, and was obliged to rely on the magnetic compass:

> In an unmapped country the conditions closely approximate to those in a dark mine, or a ship at sea, consequently the magnetic needle must play an all-important part in any instrument designed for Colonial work.
> (Bligh, 1899, p.6)

The most obvious general characteristic of an instrument intended for mine surveying is that a magnetic compass occupies a prominent and unencumbered position and is the principal means of taking bearings. The problems of magnetic disturbance and deviation encountered in a mine are, and were, obvious, and the mine surveyor's persistent use of the compass simply indicates the difficult and uncompromising conditions of his work. As one commentator put it:

241 'Lean's dial' by Troughton & Simms, mid-19th century, length 260mm. The telescope is adjusted by rack and pinion, with a clamping screw moving in a separate inner slot. The semicircle can be replaced by the detached pair of folding slit-and-window sights.

> Underground surveying . . . has been hitherto almost universally performed with the magnetic needle, that is to say with the Circumferentor. It appears so monstrous, that such a vast amount of human life, and of valuable property, should . . . be left dependent on so variable an instrument, that the statement would be almost incredible, if the fact were not so notorious. (Bourne, 1843, p.279)

Towards the end of the century, the importance of the magnetic compass was still grudgingly accepted:

> Upon this original circumferenter improvements have been made in the various mining-dials we possess, in all of which the large open compass is still preserved. This prominence of the compass does not indicate that the modern scientific mining engineer has any desire to depend upon it for taking horizontal angles, but that in close and tortuous workings it provides the nearest and often the only possible means of taking angles consistent with the extreme difficulties of observations of any kind.
> (Stanley, 1890, pp. 276–7)

Two fairly common nineteenth-century modifications of the circumferentor were Lean and Hedley's dials. A mine manager in Cornwall, Joel Lean designed a circumferentor whose two sights could be removed and replaced by an 'inverted' vertical semicircle with a telescopic sight (fig. 241). The sight, supported by a central pivot, moved around the semicircle and was equipped with a bubble level and with clamp and tangent screws. It was typical of adaptations of the circumferentor that bearings continued to be taken from the compass, while a vertical arc was added for elevations.

242 (*left*) 'Hedley's dial', by Davis & Son of Derby, c.1870, length 305mm, with a vertical semicircle and plain sights. In later examples a telescopic sight in Y bearings could be fitted to replace the plain sights. Cambridge, Whipple Museum.

243 (*right*) Three mid-19th-century prismatic compasses: left, by T. Rubergall of London, diam. 88mm, with the prism in the fore-sight; centre, by Schmalcalder, diam. 62mm; right, by Troughton & Simms, diam. 62mm. The prismatic near-sight is the more common arrangement. Cambridge, Whipple Museum.

The dial (fig. 242) invented in 1850 by John Hedley, the Inspector of Mines, became more popular than Lean's instrument, and subsequently appeared in a number of guises by different makers. Here, the ordinary circumferentor sights, instead of being attached directly to the compass box, are connected to a ring, which surrounds the compass and is attached to it by pivots at the E and W points. Thus the axis of the sights can move in altitude to take in targets out of the horizontal, and a semicircle is attached to measure altitudes. Hedley's original design had open sights, but makers soon added an optional telescope, which rested in Y bearings secured in place of the sights.

Continental mine survey instruments were generally of a rather different form, with a central and entirely unencumbered compass, and a telescopic sight and vertical circle (sometimes semicircle) carried to one side. The centre of the circle was roughly level with the compass and a counterweight was fitted to the opposite side. The compass, and sometimes the telescope, had a level, and the circular plate carrying the compass box had a degree scale, which allowed the instrument to be used as a theodolite. It was known as the 'eccentric theodolite', because the telescope was not mounted centrally.

Apart from these common forms, there were a great many specialist instruments for mine survey, which cannot be considered here in any detail. One common instrument, however, was the prismatic compass (fig. 243), invented by the maker Charles Schmalcalder and patented in 1812. A compass mounted after the manner of a marine instrument, with an engraved card, was enclosed in a circular glazed box, with fixed sights. The fore-sight was a common open sight, but incorporated into the near-sight was a reflecting prism, so that the target object and the degree scale on the card could be viewed at once and the bearing read off directly. A later variant had the prism mounted with the fore-sight. It would be wrong to classify this general-purpose device as a mine surveying instrument, but specially adapted designs were produced for mining engineers. F. W. Simms explained its more general applications:

> . . . it is particularly adapted for military surveying, or where but little more than a sketch map of the country is required. It is also very useful in filling in the detail of a map, where all the principal points have been correctly fixed by means of the theodolite; and for this purpose it has been extensively employed by the gentlemen engaged on the Ordnance survey. (Simms, 1834, p.2).

Simms had himself worked on the Ordnance Survey.

The American transit

As has been already noted, the ordinary circumferentor was widely manufactured in America, and this was symptomatic of the instrument's use in new territories, where an artificial topography had not been established by centuries of settlement. In America, however, the circumferentor was developed in the nineteenth century into an

244 Three examples of American surveying instruments, which illustrate their development from the circumferentor: in front, circumferentor by E. & G. W. Blunt, *c.*1860, length 396mm; left, surveying compass with a telescopic sight but no vertical arc, by W. & L.E. Gurley, *c.*1860, diameter 223mm; right, American transit by B. Pike & Sons, *c.*1860, diameter 155mm, with full vertical and enclosed horizontal circles. Cambridge, Whipple Museum.

245 Burt's solar campass by W. J. Young & Sons of Philadelphia, *c.*1871, length 362mm, used for finding the true meridian from the Sun and hence the magnetic variation, and also for general surveying. Cambridge, Whipple Museum.

instrument of considerable sophistication (fig. 244). First a telescopic sight was provided, with a horizontal axis carried by two A frames set on either side of the compass box. This instrument became an 'American transit' when the telescope was made to transit over the compass and was given a vertical arc. The design of the transit is usually attributed to William J. Young, though early examples were also made by Edmund Draper.

Transits came to be fitted with full vertical circles (see also fig. 181) and, ultimately, a horizontal circle was provided in addition to the compass. Instruments of this type could be described as transit theodolites with large azimuth compasses, but by this time the circular compass was disappearing from theodolites manufactured in Europe.

The natural complement to the American transit, whose accuracy depended on some knowledge of magnetic variation, was an instrument known as the 'solar compass' (fig. 245), invented in 1835 by William Austin Burt (1792–1858) of Michigan, and first manufactured by Young. The solar compass was mounted on a platform with sights at either end like the circumferentor, in the place normally occupied by the circular compass. There were three arcs: a latitude arc used to set a polar axis, which in turn carried right ascension and declination arcs and a sighting rule with a convex lens and target plate for aligning it with the Sun. When properly adjusted for latitude and solar declination, and targeted on the Sun, the instrument located the true meridian, and a trough compass with a limited degree scale gave a measure of the local magnetic variation. The sights allowed the instrument to be used for direct surveying. Solar attachments working on the same principle were later fitted to theodolites and to transits.

In 1854 Burt wrote an account of the use of the solar compass, 'assuming a task so foreign to his habits of life' (Burt, 1909, p.iv), he said in his apologetic preface, in response to the demands of practitioners. The instrument was extensively used in public surveying work and Burt's instructions comprised a detailed manual of how such work was carried on. His recommended 'Outfit for a surveying company of six men for four months in the public service' confirms some of our ill-founded impressions of frontier life, while confounding others. Provisions were to include '3 bushels of beans' and '70 lbs. of ground coffee', but also '4 lbs. of Castile soap' (*ibid*, p.79).

246 Sounding sextant by Cary, 19th century, radius 149mm. Note the large aperture telescope and the absence of shades.

Marine surveying

The particular needs of marine or hydrographic surveying gave rise to a few specialized instruments, and most surviving examples date from the nineteenth century onwards. A coastal outline would be triangulated from shore using the customary instruments, and this survey would provide the basis for the marine survey. The latter work was concerned with soundings, currents and tidal information, and the specialist survey instruments were used to locate the position of the boat, in relation to the established outline of the shore, when the sounding or other measurement was taken.

The technique was to measure the two angles between three coastal features that figured on the shore map, which were sufficient to locate uniquely the position of the boat. The instrument commonly used to measure these angles was a 'sounding sextant' (fig. 246), similar in principle and design to the ordinary sextant. In use, the instrument was, of course, held horizontally; one object was viewed directly and the other after a double reflection. No shades were provided, since none was required, and the telescope had a wide field of view. The mirrors were correspondingly larger than normal, and the 'horizon' mirror was usually not accompanied by a plane glass within the brass mount. The direct sight was made over the mirror, and the inaccuracy caused by the interposition of the edge of the mount between the two targets was judged acceptable within the overall errors of the technique.

247 (*left*) Double sounding sextant by Cary, *c*.1860, diameter 91mm. Two sets of optics are enclosed within the oxidized brass case, with two corresponding index arms and scales on either side, allowing the two angles between three objects to be taken in a single operation. Cambridge, Whipple Museum.

248 (*right*) Station pointer by Cary, mid-19th century, circle diameter 123mm. The moveable arms have clamp and tangent screws with vernier adjustment to 1 minute of arc. Each arm can be fitted with an extension piece. Cambridge, Whipple Museum.

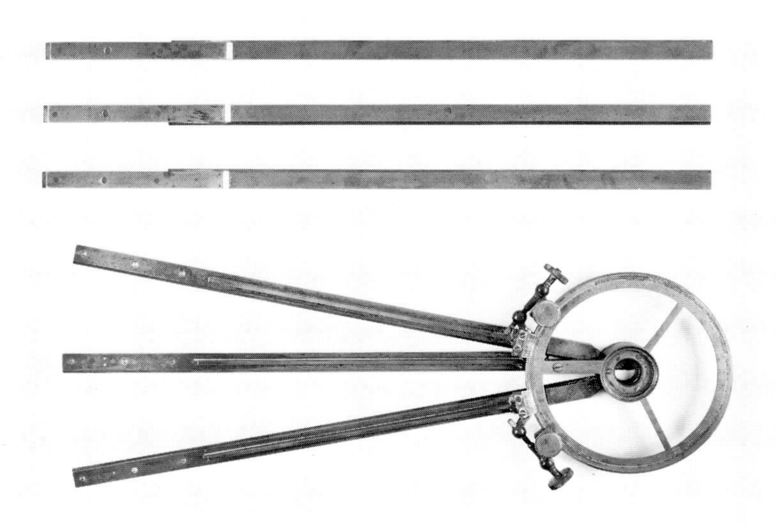

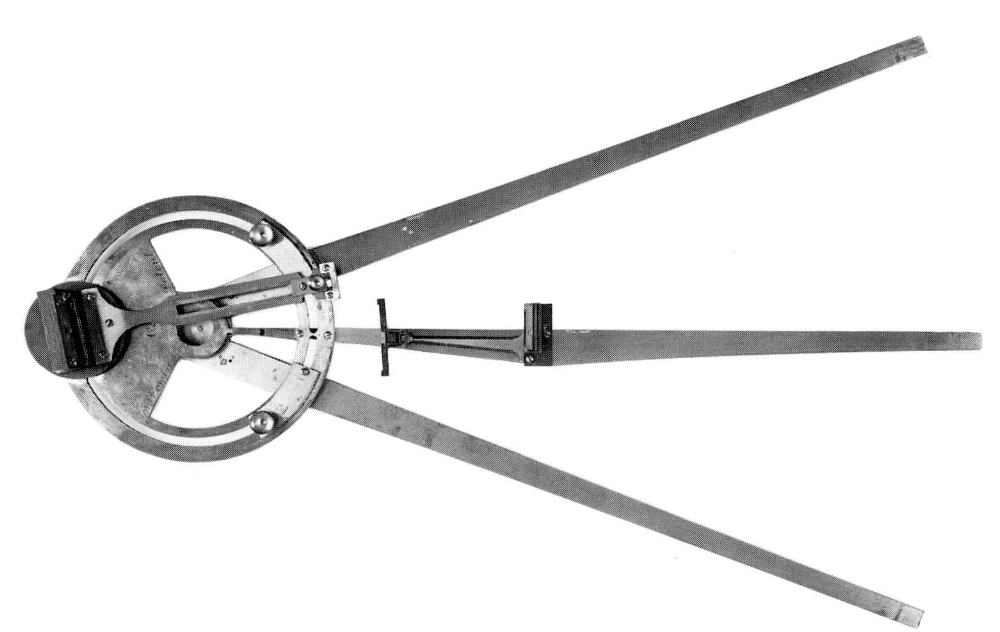

249 'McCombie's position finder', unsigned but by Hughes & Son, who were sole makers and agents (Hughes & Son, 1904), *c.*1900 A combined sounding sextant and station pointer. Cambridge, Whipple Museum.

With the ordinary sounding sextant, separate operations were required to measure the two angles. There were, however, some designs of 'double sounding sextant', or simply 'double sextant' (fig. 247), incorporating two sets of arcs, index arms and mirrors. These allowed both angles to be taken at once, by bringing three targets into coincidence, the central one being viewed directly, and those on either side by double reflection. Other more primitive and therefore cheaper instruments were also designed, such as the 'Paget angle sextant' (fig. 250), which worked on the sextant principle without the sextant's customary, but here inappropriate, accuracy. The standard designs of reflecting circle were also, of course, easily adapted to hydrographic survey.

Once the necessary angles had been measured, the positions could be located on the chart using a special form of protractor known as a 'station pointer' (fig. 248), the invention of which is generally attributed to the British hydrographer Murdoch McKenzie (d.1797). A divided circle was equipped with three radial arms extending well beyond its circumference, and the middle arm was fixed at zero. Those on either side were pivoted centrally and had clamping screws, and often tangent screws for fine adjustment. They could thus be opened to the two measured angles, and with the bevelled edges of the arms positioned on the three targets, the centre of the circle located the original position of the boat.

It was even possible to combine the station pointer with the double sounding sextant, in an instrument such as the 'McCombie position finder' (fig. 249), where the addition of sights and mirrors to the station pointer allowed its arms to be aligned directly with the targets. The instrument could then be transferred directly on to the chart. Examples of this instrument may or may not have the circle divided. With a degree scale, it could be used as a station pointer in the ordinary way, but if the sights were used, there was no need to measure the angles. Other such combinations were given the names 'Navigraph' (a development of the Paget angle sextant) and 'Goneograph', which was described in Lecky's '*Wrinkles*'.

250 A simple form of sounding sextant, known as the 'Paget angle sextant', by H. Hughes & Son Ltd, *c.*1900, diameter 114mm, in its fitted box. The handle screws into the base. This example has no telescope, but they were also available with telescopic sights. Cambridge, Whipple Museum.

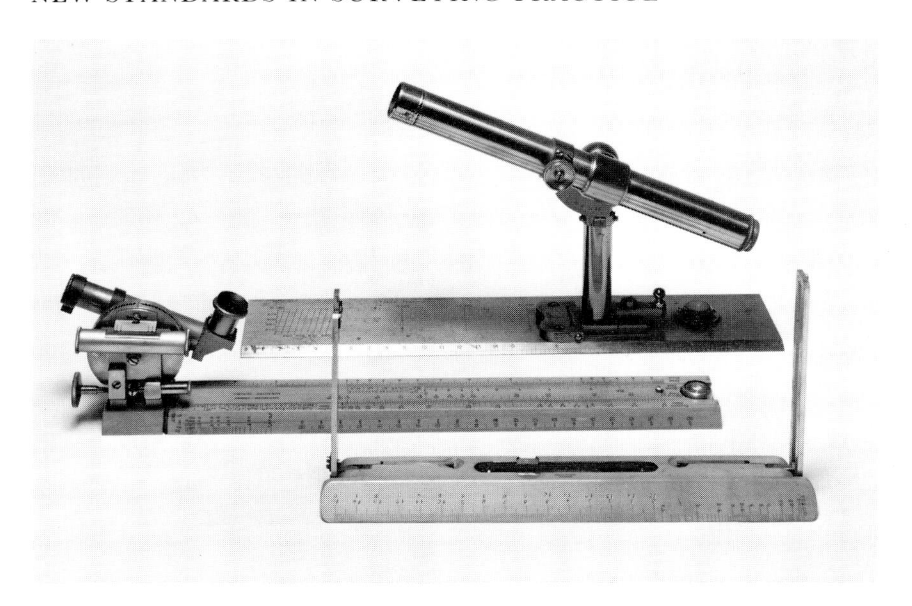

251 Three plane table alidades: in front, by Secretan, *c.*1900, length 234mm, in boxwood, with plain sights; centre, by Tavernier-Gravet of Paris, c.1915, length 328mm, with telescopic sight, vertical circle and slide-rule; back, unsigned, *c.*1900, length 350mm, with a telescopic sight and very small vertical arc. Cambridge, Whipple Museum.

Elaboration of the transit theodolite

Once it was clear that a fairly standard design of theodolite had evolved, the makers again tried to introduce more complex variants. The 'tacheometer', for example, was a theodolite equipped to measure, usually by some form of eyepiece micrometer, the apparent size of a graduated staff, and to yield a direct measure of distance. There were a number of different forms in the late nineteenth century. The 'omnimeter' was a very accurate theodolite, with readings from the vertical circle taken by a microscope, so as to distinguish the angle between the upper and lower targets on the distant staff. The 'curve ranger', invented by Sir Harley Hugh Dalrimple-Hay (1861–1940) and patented in 1886, was used specifically in railway survey. An additional adjustment to the horizontal circle allowed curves of different radii to be set out, without reference to the usual tables.

Thus, for surveying instruments, the century ended with a fair degree of standardization, around the general patterns of the transit theodolite and the dumpy level, but also with evidence that makers were still prepared to compete over more complex designs. It was only in the nineteenth century that commonplace surveying had finally come to terms with the old aspirations of the geometers, and the name of its principal tool, the 'theodolite', had come to refer unequivocally to an altazimuth instrument, modelled so long before on instruments of astronomy.

14

A Practical Postscript

IN RELATION TO THE TECHNOLOGY OF THE divided circle, our three disciplines have followed similar courses in the twentieth century. This development seals and concludes the unity between them, founded on their common geometry. Each in its turn diversified into new methods, based on new techniques, and most theoretical and instrumental innovation became concentrated on alternative technologies. The divided circle lost much of its importance for scientific development, but continued to occupy the ground where it began: it could still supply a number of practical necessities.

This process occurred in the different disciplines in the order in which they were first profoundly influenced by geometry. Thus it appeared initially in astronomy with the development of astrophysics; indeed the research emphasis in astronomy had already changed by the early years of the century. Navigation was transformed by radio signals from shore and later from satellites. Finally, the geodetic side of surveying embraced satellite technology, and the practical side came to incorporate such techniques as photogrammetry, distance measurement by electromagnetic ranging and levelling by lasers.

Astronomy

There is little to say about the divided circle in twentieth-century astronomy. Meridian work was obliged to continue even after research interests had largely moved elsewhere; it remained a practical necessity, fundamental to other measuring techniques and basic to general timekeeping. New instruments were built as necessary. At Greenwich, for example, it was agreed to build a new transit circle in 1931, and the instrument was completed in 1936. The makers were Cooke, Troughton and Simms Ltd, a company formed in 1922 by an amalgamation between Troughton and Simms and the workshop founded by Thomas Cooke (1807–68).

Much positional work, however, came to be based on photography, and even time became more accurately determined, first by quartz and then by atomic clocks. An international conference in 1967 agreed to re-define the second of time – no longer a subdivision of the astronomical day – in terms of the characteristic behaviour of an electron in the caesium atom. There are of course practical reasons why time, however defined, has to keep pace with the rotation of the Earth. Thus atomic time is adjusted as necessary, in steps of whole or 'leap' seconds, to accommodate the less accurate timekeeping of the Earth.

Navigation

The development of lighter alloys for sextant frames continued into the twentieth century, but the only change of principle may have been the replacement of the clamp and tangent screws. A disadvantage of the ordinary tangent screw is the danger of reaching the end of its travel before the observation is complete, and a solution was to arrange an 'endless tangent screw' engaging teeth around the edge of the limb. From the 1920s the vernier came to be replaced by the graduated drumhead of a micrometer screw, which might itself be fitted in turn with a small vernier.

The ordinary mechanical log – especially as designed and manufactured by Walker – continued in use for at least the first half of the century, but was supplemented by new designs. The 'Chernikeef', for example, made in the 1930s by the Electrical Submerged Log Company of London, set the rotator in a tube attached to the ship's hull and connected it to a dial on the bridge. The arrangement of the 'Pilometer' was similar, but it worked on a different principle – by measuring the increase in the forward pressure of sea-water caused

252 (*left*) The 'navy pattern' of standard compass as illustrated in a 1904 Hughes & Son catalogue. The Thomson binnacle could be fitted with either Thomson's dry-card compass or a liquid compass.

253 (*right*) Simple theodolite by Stanley, *c.*1900, width 254mm. An instrument for measuring azimuths only, for which the name 'simple theodolite' was used by Stanley in his standard work on surveying instruments (Stanley, 1890, pp. 256–8).

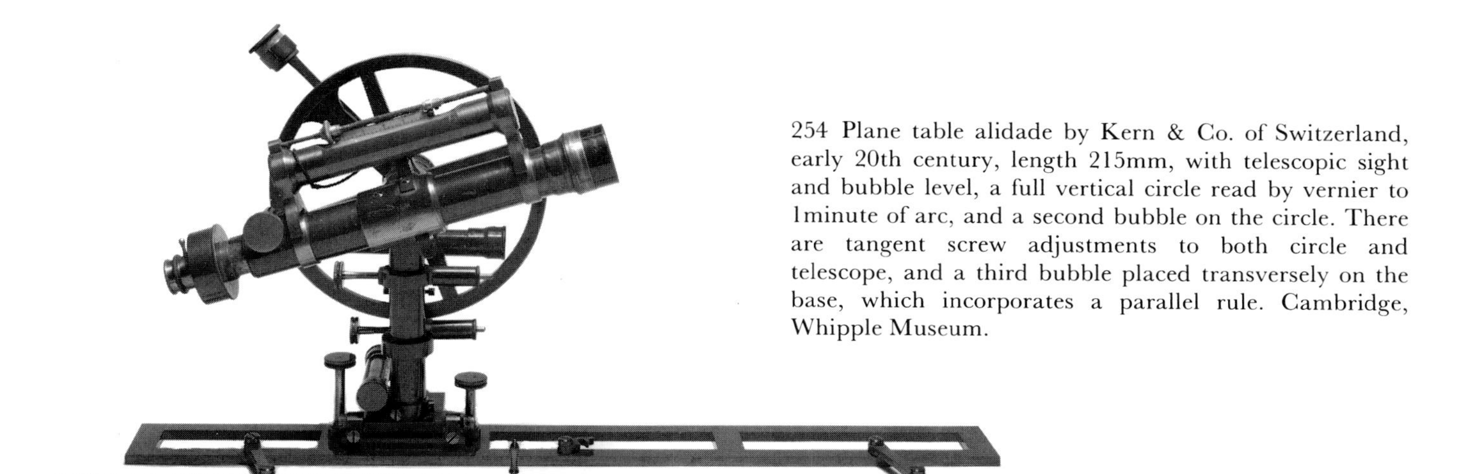

254 Plane table alidade by Kern & Co. of Switzerland, early 20th century, length 215mm, with telescopic sight and bubble level, a full vertical circle read by vernier to 1minute of arc, and a second bubble on the circle. There are tangent screw adjustments to both circle and telescope, and a third bubble placed transversely on the base, which incorporates a parallel rule. Cambridge, Whipple Museum.

by the ship's motion. These three types of log were described as being in general use in the 1957 edition of Lecky's '*Wrinkles*'.

In time, the dry-card compass gave way to the liquid-filled model, but this was still housed in a binnacle closely related to Thomson's design. In 1952 Grant and Klinkert, in their textbook *The ship's compass*, described the standard magnetic instrument as a liquid compass with a card of about 6 in diameter in a bowl of 9 in. A card of mica with a hemispherical float, a sapphire bearing and two cobalt steel magnets, was held in place by an osmium-iridium point. The Flinders bar and correcting spheres were still mounted outside the binnacle, while within were the various permanent magnets and an electric light for illuminating the card from underneath.

By then, however, an altogether different instrument was available – the gyro-compass, based on the directional property of a spinning gyroscope mounted so as to align itself with the axis of the Earth's rotation. Yet this did not end the development of the magnetic compass, since the gyro-compass was susceptible to malfunction or to power failure. The gyro could readily operate 'repeaters' to reproduce its reading at different positions on the ship, and the same facility was developed for the magnetic compass. The 'projection compass' was a simple form, where a lamp was used to project an image to the steering position. A 'transmitting compass' transferred signals from electrical contacts or from selenium cells activated by light admitted through apertures in the card.

The oldest instrument of all, the lead and line, was replaced by echo-sounding, developed during the 1920s, and the compass could be supplemented by a 'radio compass' or 'direction finder', which detected the direction of a radio beacon. Without a compass, such a direction could be found only relative to the ship's head, but an absolute determination was possible by integrating readings from other beacons. Distances cound be found to relatively nearby beacons transmitting simultaneous radio and sound signals. The development of electronic techniques led to such complete position finding systems as 'Consol' (originating in a German system of World War Two), 'Lorcan' (developed by the United States during the same period), 'Decca' (first operational in 1946) and 'Omega' (1968). The late 1960s also saw the first extension of satellite navigation beyond its military use.

255 Miner's dial by E. R. Watts & Son, 1931, diameter 145mm. A late example of a miner's dial with vertical and horizontal circles, but retaining the characteristic feature of a prominent and unencumbered compass. Cambridge, Whipple Museum.

In spite of these revolutionary approaches to navigation, the traditional astronomical techniques survive and ships continue to carry sextants. Electronic navigation relies on such artificial aids as satellites and radio stations, not only susceptible to malfunction, but vulnerable in time of war. The divided circle survives in navigation, not only as a practical expedient, but as a reminder of the precarious dependence of our technological society on international goodwill.

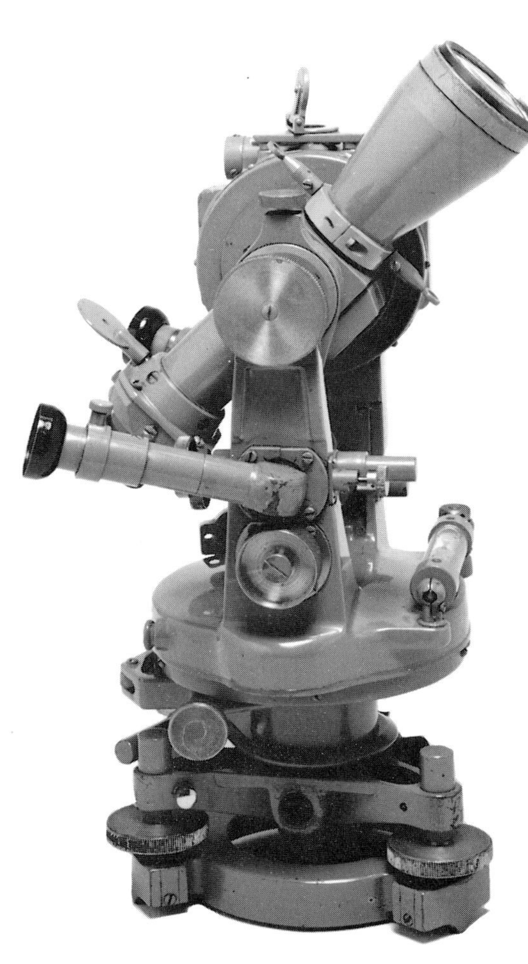

Surveying

The transit theodolite and the dumpy level have continued to provide the basic tools of the ordinary surveyor. The most obvious development in the theodolite has been the increasing enclosure of first the scales and then the other mechanical and optical components. Micrometer microscopes were used in instruments aimed at greater precision, and the use of prisms allowed microscopes – whether micrometer or reading – to be positioned at convenient angles. Eventually scales were etched on glass and illuminated from behind, and the images of the horizontal and vertical circles carried through the enclosed optical components to a single eyepiece, and there displayed together.

Large-scale operations increasingly employed techniques of photogrammetry, as a fast and economic supplement to field work; the map was constructed from measurements taken from aerial photographs. The most radical development in land-based survey has been 'electronic distance measurement' (EDM), based on timing light, microwave, laser or infra-red beams reflected from a target. In the mid-1970s the miniaturization of electronics allowed EDM instruments to be mounted on top of transit theodolites. Theodolites themselves could be 'digitized', with glass 'encoder' discs replacing divided circles and photodiodes responding to the quantity

256 (*above*) Transit theodolite by Cooke, Troughton & Simms, *c.*1942, diameter (horizontal circle) 110mm, with totally enclosed circles, read by micrometers through enclosed optics to 1 second of arc. The model was called 'the Tavistock', after the Devon venue of a conference between survey officers and manufacturers that established its specification. Cambridge, Whipple Museum.

257 (*right*) Drainage level by E. R. Watts & Son, *c.*1915, length 357mm, with a prismatic compass in the base and a graduated micrometer screw for adjusting the inclination of the telescope. Cambridge, Whipple Museum.

258 (*above*) Three levels: left, by Ertel & Sohn, mid-19th century, length 267mm; centre, by Stanley, *c.*1890, length 345mm, with a prismatic compass incorporated into the base; right, by Troughton & Simms, 1846, length 417mm. The latter two are dumpy levels, each with the characteristic transverse bubble (cf. fig. 237). Cambridge, Whipple Museum.

259 (*left*) Three levels: left, by Wild Heerbrugg of Switzerland, 1980, length 163mm; centre, by E.R. Watts & Son, *c.*1920, length 200mm; right, by M.D.S. Ltd of London, 1939, length 344mm. The instrument by Watts, which the manufacturers called a 'reflecting level', incorporates a bubble but is also a prismatic compass. Cambridge, Whipple Museum.

of transmitted light, characteristic of the measured angle. The inclusion of a micro-processor and a solid-state memory allowed an instrument to function as a complete positioning system, with computing and recording capacity. Lasers could be used for establishing alignments and gyro attachments gave absolute directions.

The transit theodolite and the dumpy level still serve many practical needs, and it is interesting that the more comprehensive modern textbooks still deal with the plane table, although the alidade now has a telescopic sight and an enclosed vertical circle read by a microscope.

Conclusion

This treatment of twentieth-century instrumentation has been sketchy in the extreme, but this is appropriate, since it represents the demise of the divided circle. For roughly two millennia its technology had epitomized scientific accuracy. It was linked on the one hand with the Euclidean geometry of the cosmos, and on the other with the practical application of mathematics to aspects of everyday life. Fashioned in brass, silver, ivory or wood, the divided circle had mediated between the theories of geometers and astronomers, and the immediate necessities of navigators and surveyors.

Its history had involved many of the leading men of science – Ptolemy, Regiomontanus, Tycho, Gemma, Hooke, Cassini, Halley, Graham, Ramsden, Fraunhofer, Airy, Kelvin. It had also depended for its widespread use on many more who are little known – Hood, Whitwell, Sutton, Chapotot, Heath, Breithaupt, Stancliffe, Pike, Morin, Gravatt, Lean, Gurley. Together their achievements included detecting the motion of the Earth, measuring the distance of the stars, discovering the size and shape of the globe, and finding the longitude.

After the astronomers had discovered new areas of research, and the mathematicians a new variety of cosmic geometry, alternative technologies were found to improve on the performance of the divided circle. The instruments are left to play two roles. Recent examples have a remaining, mundane usefulness, bereft of an inspirational link with research science. Older instruments form an astonishing material record, an evocation of a past technology that combined beauty and elegance with craftsmanship, precision and hard-headed utility.

Bibliography

D.S.B. = Gillispie, C. C., ed., *Dictionary of scientific biography* (New York, 1970–80).

Adams, G., *Geometrical and graphical essays* (London, 1791).

Adams, G., ed. W. Jones, *Geometrical and graphical essays* (London, 1797) and (London, 1803).

Adams, G., ed. W. Jones, *Lectures in natural philosophy* (London, 1799).

Agricola, G., *De re metallica libri XII* (Basel, 1556).

Alexander, A. F. O'D., 'Bradley, James', *D.S.B.*.

Anderson, R. G. W., *The mariner's astrolabe* (Edinburgh, 1972).

Andrews, J. H., *A paper landscape. The Ordnance Survey in nineteenth-century Ireland* (Oxford, 1975).

Andrews, J. H., *Plantation acres. An historical study of the Irish land surveyor and his maps* (Ulster Historical Foundation, 1985).

Apianus, P., *Quadrans Apiani astronomicus* (Ingolstadt, 1532)

Apianus, P., *Cosmographia, per Gemmam Phrysum ... restituta* (Antwerp, 1539) and (Antwerp, 1550).

Apianus, P., *Astronomicum Caesareum* (Ingolstadt, 1540).

Apianus, P., see Werner, J., *Introductio geographica ...*

Archinard, M., *Baromètres* (Geneva, 1978).

Archinard, M., *Collection de Saussure* (Geneva, 1979).

Baily, F., ed., *An account of the life of the Revd. John Flamsteed* (London, 1835).

Bedini, S. A., *Early American scientific instruments and their makers* (Washington, 1964).

Bedini, S. A., *Thinkers and tinkers. Early American men of science* (New York, 1975).

Bedini, S. A., *At the sign of the Compass and Quadrant. The life and times of Anthony Lamb* (Philadelphia, 1984).

Bennett, J. A., 'A study of Parentalia, with two unpublished letters of Sir Christopher Wren', *Annals of science*, 30 (1973), pp. 129–147.

Bennett, J. A., 'Robert Hooke as mechanic and natural philosopher', *Notes and records of the Royal Society of London*, 35 (1980), pp. 33–48.

Bennett, J. A., *The mathematical science of Christopher Wren* (Cambridge, 1983).

Bennett, J. A., *Whipple Museum of the History of Science. Catalogue 3, Astronomy and navigation* (Cambridge, 1983, A)

Bennett, J. A., *Science at the Great Exhibition* (Cambridge, 1983, B).

Bennett, J. A., 'Instrument makers and the "Decline of science in England": the effect of institutional change on the elite makers of the early nineteenth century', in De Clercq, 1985.

Bennett, J. A., 'The longitude and the new science', *Vistas in astronomy*, 28 (1985), pp. 219–25.

Bennett, J. A. 'The mechanics' philosophy and the mechanical philosophy', *History of science*, 24 (1986), pp. 1–28.

Bennett, J. A., and Brown, O., *The compleat surveyor* (Cambridge, 1982).

Bion, N., *Traité de la construction et des principaux usages des instrumens de mathematique* (Paris, 1709).

Bion, N., trans. Stone, E., *The construction and principal uses of mathematical instruments* (London, 1723) and (London, 1758).

Bird, J., *The method of dividing astronomical instruments* (London, 1767).

Bird, J., *The method of constructing mural quadrants* (London, 1768).

Blagrave, J., *Baculum familliare ... the familiar staffe* (London, 1590).

Blundeville, T., *His exercises* (London, 1594).

Bond, H., *The longitude found* (London, 1676).

Bonelli, M. L. R., *Il Museo Di Storia Della Scienza a Firenze* (Milan, 1968).

Borough, W., *A discours of the variation of the cumpas, or magneticall needle* (London, 1581).

Bos, H. J. M., *et al*, eds, *Studies on Christiaan Huygens* (Lisse, 1980).

Bourne, W., *A regiment for the sea* (London, 1574).

Bourne, W., ed. Taylor, E. G. R., *A regiment for the sea, and other writings on navigation* (Cambridge, 1963).

Brachner, A., *et al*, *G. F. Brander, 1713–1783. Wissenschaftliche Instrumente aus seiner Werkstatt* (Munich, 1983).

Brachner, A., 'German nineteenth-century scientific instrument makers', in De Clercq, 1985.

Bradley, J., 'A letter to the Rt. Hon. George, Earl of Macclesfield, concerning an apparent motion in some of the fixed stars'. *Philosophical transactions*, 45 (1748), pp. 1–43.

Brahe, T., *Astronomiae instauratae mechanica* (Wandesburgi, 1598).

Brahe, T., trans. and ed. Raeder, H., *et al*, *Tycho Brahe's description of his instruments and scientific work* (København, 1946).

Bramer, B., *Bericht zu Jobsten Burgi seligen geometrischen triangular Instruments* (Kassel, 1648).

Brenni, P., 'Italian scientific instrument makers of the nineteenth century and their instruments', in De Clercq, 1985.

Brewington, M. V., *The Peabody Museum collection of navigation instruments* (Salem, 1963).

Brown, J., *Mathematical instrument-makers in the Grocers' Company 1688–1800* (London, 1979).

Brown, O., *Whipple Museum of the History of Science. Catalogue 1, Surveying.* (Cambridge, 1982).

Burt, W. A., *A key to the solar compass and surveyor's companion* (8th edn, New York, 1909).

Chambers, G. F., *A handbook of descriptive and practical astronomy* (4th edn, Oxford, 1890).

Chapman, A., 'The accuracy of angular measuring instruments used in astronomy between 1500 and 1850', *Journal for the history of astronomy*, 14 (1983), pp. 133–7.

Close, C., *The early years of the Ordnance Survey* (2nd edn, 1969).

Cole, B., *A new sea quadrant* (London, 1748).

Cortés, M., trans. Eden, R., *The arte of navigation* (London, 1579).

Cotter, C. H., *A history of nautical astronomy* (London, 1968).

Cuningham, W., *The cosmographical glasse* (London, 1559).

Danfrie, P., *Déclaration de l'usage du graphomètre* (Paris, 1597).

Daumas, M., *Les instruments scientifiques aux XIIe et XVIIIe siècles* (Paris, 1953).

Daumas, M., trans. Holbrook, M., *Scientific instruments of the seventeenth and eighteenth centuries and their makers* (London, 1972).

Davis, J., *The seamans secrets* (London, 1595).

Davis, R. E., *et al*, *Surveying theory and practice* (6th edn, New York, 1981).

De Clercq, P. R., ed., *Nineteenth-century scientific instruments and their makers* (Leiden and Amsterdam, 1985).

De Clercq, P. R., 'The scientific instrument-making industry in the Netherlands in the nineteenth century', in De Clercq, 1985.

Dee, J., 'The compendius rehersal ...', in Hearne, T., ed., *Johannis, confratis & monachi Glastoniensis, chronica ...*, vol. 2 (Oxford, 1726).

De Smet, A., 'Mercator à Louvain (1530–1552)', in *Gerhard Mercator 1512–1594. Duisburger Forschungen*, band 6 (1962), pp. 28–90.

Dewhirst, D. W., 'Meridian astronomy in the private and university observatories of the United Kingdom: rise and fall', *Vistas in astronomy*, 28 (1985), pp. 147–158.

Digges, L., *A booke named Tectonicon* (London, 1556).

Digges, L., ed. Digges, T., *A geometrical practise, named Pantometria* (London, 1571) and (London, 1591).

Digges, T., *A perfit description of the caelestiall orbes* (London, 1576).

Dizer, M., ed., *International symposium on the observatories in Islam* (Istanbul, 1980).

Donnelly, M. C., *A short history of observatories* (Oregon, 1973).

Dreier, F. A., ed., *Winkelmessinstrumente vom 16. bis zum frühen 19. Jahrhundert* (Berlin, 1979).

Dreyer, J. L. E., *Tycho Brahe. A picture of scientific life and work in the sixteenth century* (London, 1890).

Dreyer, J. L. E., *A history of astronomy from Thales to Kepler* (2nd edn, New York, 1953).

Fine, O., *De mundi sphaera, sive cosmographia* (Paris, 1542).

Fine, O., *De re & praxi geometrica* (Paris, 1586).

Fine, O., *Opere* (Paris, 1587).

Flamsteed, J., *Historia coelestis Britannicae* (London, 1725).

Folkerts, M., 'Werner, Johann', *D.S.B.*

Forbes, E. G., *The birth of navigational science* (London, 1973).

Forbes, E. G., *Greenwich Observatory. Vol. 1, Origins and early history,* (London, 1975).

Gallucci, G. P., *Della fabrica et uso di diversi stromenti di astronomia et cosmografia* (Venice, 1598).

Gardiner, W., *Practical surveying improved* (London, 1737).

Gellibrand, H., *A discourse mathematical on the variation of the magneticall needle* (London, 1635).

Gemma Frisius, *De radio astronomico & geometrico liber* (Paris, 1558).

Gibson, R., *A treatise of practical surveying* (4th edn, Dublin, 1777).

Gingerich, O., ed., *The general history of astronomy, vol. 4. Astrophysics and twentieth-century astronomy to 1950: part A* (Cambridge, 1984).

Gould, R. T., *The marine chronometer, its history and development* (London, 1923).

Gray, J., *The art of land-measuring* (Glasgow, 1757).

Gunter, E., *De radio et sectore* (London, 1623).

Gunther, R. T., *The astrolabes of the world* (Oxford, 1932).

Gurley, W. & L. E., *A manual of the principal instruments used in American engineering and surveying* (19th edn, Troy, New York, 1873).

Guyot, E., *Historie de la détermination des longitudes* (La Chaux-de-Fonds, 1955).

Hamann, G., ed., *Regiomontanus-studien* (Vienna, 1950).

Hammond, J., *The practical surveyor* (London, 1725).

Hammond, J., ed. Warner, S., *The practical surveyor* (London, 1731) and (London, 1750).

Haskoll, W. D., *Land and marine surveying* (London, 1868).

Hellman, C. D., 'Brahe, Tycho', *D.S.B.*

Hellman, C. D., and Swerdlow, N. M., 'Peurbach, Georg', *D.S.B.*

Herrmann, D. B., 'Wilhelm IV, Landgrave of Hesse', *D.S.B.*

Hevelius, J., *Machina coelestis* (Danzig, 1673, 9).

Hewson, J. B., *A history of the practice of navigation* (Glasgow, 1951).

Hinks, A. R., *Maps and survey* (Cambridge, 1913) and later editions.

Hooke, R., *Animadversions on the first part of the Machina Coelestis of ... Johannes Hevelius* (London, 1674).

Hooke, R., ed. Robinson, H. W., and Adams, W., *The diary of Robert Hooke* (London, 1935).

Hopton, A., *Baculum geodeticum ... or the geodeticall staffe* (London, 1610).

Hopton, A., *Speculum topographicum: or the topographicall glasse* (London, 1611).

Horrebow, P., *Basis astronomiae* (Copenhagen, 1741).

Howse, D., *Greenwich Observatory. Vol. 3, The buildings and instruments* (London, 1975).

Howse, D., *Greenwich time and the discovery of longitude* (Oxford, 1980).

Howse, D., 'The Greenwich list of observatories: a world list of astronomical observatories, instruments and clocks, 1670–1850', *Journal for the history of astronomy*, 17 (1986).

Hughes, Henry, & Son, Ltd, [Catalogue] (London, 1904).

Inwards, R., *William Ford Stanley. His life and works* (London, 1911).

Johnston, S. A., Willmoth, F. H., and Bennett, J. A., *The grounde of artes. Mathematical books of 16th-century England* (Cambridge, 1985).

Kari-Niazov, T. N., 'Ulugh Beg', *D.S.B.*

Kiely, E. R., *Surveying instruments, their history and classroom use* (New York, 1947).

King, H. C., *The history of the telescope* (London, 1955).

King, H. C., *Geared to the stars. The evolution of planetariums, orreries and astronomical clocks* (Bristol, 1978).

Kirchvogel, P. A., Wilhelm IV, Tycho Brahe, and Eberhard Baldewein – the missing instruments of the Kassel Observatory', *Vistas in astronomy,* 9 (1967), pp. 109–21.

Kish, G., 'Apian, Peter', 'Dudley, Robert', 'Gemma Frisius, Reiner', 'Mercator, Gerardus', 'Waldseemüller, Martin', *D.S.B.*

Kissam, P., *Surveying for civil engineers* (New York, 1981).

Kopal, Z., Römer, Ole Christensen', *D.S.B.*

La Lande, J. Le F. de, *Astronomie* (Paris, 1764) and (Paris, 1792).

Lamb, U., trans., *A navigator's universe. The Libro de Cosmographia of 1538 by Pedro de Medina* (Chicago and London, 1972).

Lancaster-Jones, E., *Catalogue of the collections in the Science Museum ... Geodesy and surveying* (London, 1925).

Lecky, S. T. S., *'Wrinkles' in practical navigation* (London, 1881) and subsequent editions to (London, 1925).

Leybourn, W., *The compleat surveyor* (London, 1653) and later editions.

Lindberg, D. C., ed., *Science in the Middle Ages* (Chicago, 1978).

L'industrie française des instruments de précision, 1901–1902. Catalogue (Paris, 1901, 1902), reprint (Paris, 1980).

Lockyer, J. N., *Stargazing: past and present* (London, 1878).

López de Azcona, 'Cortés de Albacar, Martin', 'Medina, Pedro de', Nuñez Salaciense, Pedro', *D.S.B.*

McKeon, R. M., 'Les debuts de l'astronomie de précision', *Physis*, 13 (1971), pp. 225–88; 14 (1972), pp. 221–42.

MacPike, E. F., *Hevelius, Flamsteed and Halley* (London, 1937).

Maddison, F. R., 'Early astronomical and mathematical instruments', *History of science*, 2 (1963), pp. 17–50.

Maddison, F. R., 'Medieval scientific instruments and the development of navigational instruments in the XVth and XVIth centuries', *Agrupamento de estudos de cartografia antiga*, 30 (Coimbra, 1969).

Marquet, F., *Histoire générale de la navigation du XVe an XXe siècle* (Paris, 1931).

Masotti, A., 'Tartaglia, Niccolo', *D.S.B.*

May, W. E., *A history of marine navigation* (Hedley-on-Thames, 1973).

Medina, P. de, *Arte de navegar* (1545, reprinted Madrid, 1945).

Medina, P. de, *Regimiento de navegacion* (1563, reprinted Madrid, 1964).

Mercer, V., *John Arnold & Son, chronometer makers* (London, 1972).

Mercer, V., *The life and letters of Edward John Dent* (London, 1977).

Mercer, V., *The Frodshams* (London, 1981).

Michel, H., *Instruments des sciences dans l'art et l'histoire* (Rhode-Saint-Genèse, Belgium 1966).

Michel, H., *Traité de l'astrolabe* (Paris, 1976).

Minow, H., ed., *Historische vermessungsinstrumente* (Wiesbaden, 1982).

Moskowitz, S., *Historical Technology Inc.,* Catalogue series.

Mudge, W., *et al*, *An account of the operations carried on for accomplishing a trigonometrical survey of England and Wales* (London, 1799–1811).

Museo di Storia della Scienza Firenze, *Catalogo degli strumenti* (Florence, 1954).

Nasr, S. H., 'Al-Tusi, Muhammad ibn Muhammad ibn al-Hasan', *D.S.B.*

Nasr, S. H., *Science and civilization in Islam* (Cambridge, Mass., 1968).

Nasr, S. H., *Islamic science. An illustrated history* (London, 1976).

National Maritime Museum, *An inventory of the navigation and astronomy collections* (London, 1970, 3, 8, 1982).

National Maritime museum, Department of Navigation and Astronomy, *The planispheric astrolabe* (London, 1976).

Negretti & Zambra, *Encyclopaedic illustrated and descriptive reference catalogue* (London, 1886).

Neugebauer, O., 'Ancient mathematics and astronomy', in Singer, 1954, 1, pp. 785–803.

Neugebauer, O., *The exact sciences in antiquity* (2nd edn, Providence, Rhode Island, 1957).

Newton, I., 'A true copy of a paper found, in the hand writing of Sir Isaac Newton, among the papers of the late Dr Halley, containing a description of an instrument for observing the Moon's distance from the fixt stars at sea', *Philosophical transactions,* 42 (1742), pp. 155–6.

Nordon, J., *The surveyor's dialogue* (London, 1607).

Norman, R., *The newe attractive* (London, 1581).

North, J. D., 'Hevelius, Johannes', 'Richard of Wallingford', *D.S.B.*

North, J. D., *Richard of Wallingford* (Oxford, 1976).

Nuñez, P., *Tratado da sphera* (Lisbon, 1537, reprinted Munich, 1915).

O'Donoghue, Y., *William Roy, 1726–1790. Pioneer of the Ordnance Survey* (London, 1977).

Osley, A. S., *Mercator* (London, 1969).

Payen, J., 'La construction des instruments scientifiques en France au XIXe siècle', in De Clercq, 1985.

Pearson, W., *An introduction to practical astronomy*, vol 2 (London, 1829).

Penther, J. F., *Praxis geometriae* (Augsburg, 1723).

Picard, J., *Traité du nivellement* (nouvelle edition, Paris, 1780).

Pike, B., Jr, *Pike's illustrated descriptive catalogue of optical, mathematical, and philosophical instruments* (New York, 1848).

Pipping, G., *The chamber of physics* (Stockholm, 1977).

Plotkin, H., 'Molyneux, Samuel', *D.S.B.*

Poulle, E., 'Fine, Oronce', *D.S.B.*

Price, D. J., 'Precision instruments: to 1500', 'The manufacture of scientific instruments from c. 1500 to c. 1700', Singer, 1954, 3, pp. 582–619, 620–47.

Ptolemy, C., trans. Toomer, G. J., *Ptolemy's Almagest* (London, 1984).

Puissant, L., *Traité de géodésie* (Paris, 1805).

Puissant, L., *Traité de topographie, d'arpentage et de nivellement* (Paris, 1807).

Quill, H., *John Harrison, the man who found longitude* (London, 1966).

Ramsden, J., *Description of an engine for dividing mathematical instruments* (London, 1777).

Randier, J., *L'instrument de marine* (Paris, 1978).

Randier, J., trans. Powell J. E., *Marine navigation instruments* (London, 1980).

Rathborne, A., *The surveyor* (London, 1616).

Recorde, R., *The castle of knowledge* (London, 1556).

Regiomontanus, J., *Scripta de torqueto* (Nuremberg, 1544).

Repsold, J. A., *Zur geschichte der astronomischen messwerkzeuge van Purbach bis Reichenbach* (Leipzig, 1908).

Richeson, A. W., *English land measuring to 1800: instruments and practices* (Cambridge, Mass., 1966).

Roche, J. J., 'The radius astronomicus in England', *Annals of science*, 38 (1981), pp. 1–32.

Rosen, E., 'Regiomontanus, Johannes', *D.S.B.*

Roy, W., 'An account of the trigonometrical operation, whereby the distance between the meridians of the Royal Observatories of Greenwich and Paris has been determined', *Philosophical transactions*, 80 (1790), pp. 111–270.

Saunders, H. M., *All the astrolabes* (Oxford, 1984).

Sayili, A., *The observatory in Islam* (Ankara, 1960).

Seymour, W. A., ed., *A history of the Ordnance Survey* (Folkestone, Kent, 1980).

Scott, D. D., *et al*, *The evolution of mine-surveying instruments* (New York, 1902).

Short, J., 'Description and uses of the equatorial telescope or portable observatory', *Philosophical transactions*, 46 (1749–50), pp. 241–6.

Simms, F. W., *Treatise on the principal mathematical instruments employed in surveying, levelling and astronomy* (London, 1834).

Simms, W., *The achromatic telescope and its various mountings, especially the equatorial* (London, 1852).

Singer, C., *et al*, eds, *A history of technology* (Oxford, 1954–84).

Skempton, A. W., & Brown, J., 'John and Edward Troughton, mathematical instrument makers', *Notes and records of the Royal Society of London*, 27 (1972–3), pp. 233–62.

Smart, C. E., *The makers of surveying instruments in America since 1700* (Troy, New York, 1962).

Smith, C., *The description ... of a new instrument, or sea quadrant* (London, no date).

Smith, R. *A compleat system of opticks* (Cambridge, 1738).

Stanley, W. F., *Surveying and levelling instruments* (London, 1890).

Stimson, A. N., 'Some Board of Longitude instruments in the nineteenth century', in De Clercq, 1985.

Stimson, A. N., and Daniel, C. St J. H., *The cross staff. Historical development and modern use* (London, 1977).

Stoeffler, J., *Elucidatio fabricae ususque astrolabii* (Oppenheim, 1524).

Stott, C., 'The Greenwich meridian instruments', *Vistas in astronomy*, 28 (1985), pp. 133–146.

Symonds, R. W., *Thomas Tompion, his life and work* (London, 1951).

Tarozzi, G., ed., *Gli strumenti nella storia e nella filosofia della scienza* (Bologna, 1983).

Tartaglia, N., *Quesiti, et inventioni diverse* (Venice, 1546).

Tartaglia, N., *Nova scientia* (Venice, 1537) and (Venice, 1550).

Taton, R., 'Cassini, Gian Domenico', *D.S.B.*

Taton, R., ed., *A general history of the sciences, 1, Ancient and medieval science* (London, 1957).

Taylor, E. G. R., *Tudor geography 1485–1583* (London, 1930).

Taylor, E. G. R., *Late Tudor and early Stuart geography 1583–1650* (London, 1934).

Taylor, E. G. R., 'Cartography, survey, and navigation 1400–1750', in Singer, 1954.

Taylor, E. G. R., *The mathematical practitioners of Tudor and Stuart England* (Cambridge, 1954).

Taylor, E. G. R., *The haven-finding art* (London, 1956).

Taylor, E. G. R., *The mathematical practitioners of Hanoverian England* (Cambridge, 1966).

Taylor, E. G. R., and Richey, M. W., *The geometrical seaman* (London, 1962).

Taylor, W. B. S., *History of the University of Dublin* (London, 1845).

Thompson, S. P., *The life of William Thomson* (London, 1910).

Thoren, V. E., 'New light on Tycho's instruments', *Journal for the history of astronomy*, 4 (1973), pp. 25–45.

Thorndike, L., 'Franco de Polonia and the turquet', *Isis*, 36 (1845), pp. 6–7.

Todhunter, I., *A history of the mathematical theories of attraction and the figure of the Earth* (London, 1873).

Turner, A. J., *The Time Museum, volume 1. Time measuring instruments, part 1, astrolabes, astrolabe related instruments* (Rockford, Illinois, 1985).

Turner, G. L.' E., *Antique scientific instruments* (Poole, Dorset, 1980).

Turner, G. L.' E., *Nineteenth-century scientific instruments* (London, 1983).

Turner, G. L.' E., 'Mathematical instrument-making in London in the sixteenth century', in Tyacke, 1983.

Turner, G. L.' E., and Levere, T. H., *Martinus Van Marum, life and work. Volume 4, Van Marum's scientific instruments in Teyler's Museum* (Leyden, 1973).

Tyacke, S., ed., *English map-making, 1500–1600* (London, 1983).

Van Ortroy, F., *Bio-bibliographie de Gemma Frisius ... de son fils Corneille et de ses neveux les Arsenius*, Académie Royale de Belgique, *Mémoires*, 11 (1920).

Vince, S., *A treatise on practical astronomy* (Cambridge, 1790).

Ward, F. A. B., *A catalogue of European scientific instruments in the Department of Medieval and Later Antiquities of the British Museum* (London, 1981).

Warner, G. F., ed., *The voyage of Robert Dudley ... to the West Indies, 1594–1595* (London, 1899).

Wartnaby, J., *Surveying. Instruments and methods* (London, 1968).

Waters, D. W., *The art of navigation in England in Elizabethan and early Stuart times* (London, 1958).

Wattenberg, D., *Peter Apianus und sein Astronomicum Caesareum.* (Leipzig, 1967).

Weiser, R. von, *Die Cosmographiae Introductio des Martin Waldseemüller* (Strassburg, 1907).

Werner, J., *Introductio geographica Petri Apiani in Doctissimas Verneri annotationes* (Ingolstadt, 1533).

Wesley, W. G., 'The accuracy of Tycho Brahe's instruments', *Journal for the history of astronomy*, 9 (1978), pp. 42–53.

Westfall, R. S. , *Never at rest. A biography of Isaac Newton* (Cambridge, 1980).

Wheatland, D. P., *The apparatus of science at Harvard 1765–1800* (Cambridge, Mass., 1968).

Wolf, A., *A history of science, technology, and philosophy in the eighteenth century* (2nd edn, London, 1952).

Wolf, C., *Histoire de l'Observatoire de Paris de sa fondation à 1793* (Paris, 1902).

Woolf, H., *The transits of Venus* (Princeton, 1959).

Wyld, S., *The practical surveyor* (London, 1725) and (London, 1734).

Wynter, H., and Turner, A. J., *Scientific instruments* (London, 1975).

Zinner, E., *Deutsches und Niederländische astronomische Instrumente des 11.–18. Jahrhunderts* (Munich, 1956).

Zinner, E., *Leben und wirken der Joh. Müller von Königsberg genannt Regiomontanus* (Osnabrück, 1968).

Index of Makers

Index of Technical Terms

Note Technical terms are defined as they are introduced in the text. The following index gives summary definitions where appropriate and refers the reader to the text for full explanations.

General Index

11-10-93

Pete Van Evera